Farm animal proteomics

Farm animal proteomics

Proceedings of the 3rd Managing Committee Meeting and
2nd Meeting of Working Groups 1, 2 & 3 of COST Action FA1002

Vilamoura, Algarve, Portugal

12-13 April 2012

edited by:
Pedro Rodrigues
David Eckersall
André de Almeida

Wageningen Academic
P u b l i s h e r s

ISBN: 978-90-8686-195-8
e-ISBN: 978-90-8686-751-6
DOI: 10.3920/978-90-8686-751-6

Cover drawing by Simão Mateus

First published, 2012

© **Wageningen Academic Publishers**
The Netherlands, 2012

The Cost Organisation

COST – the acronym for European Cooperation in Science and Technology – is the oldest and widest European intergovernmental network for cooperation in research. Established by the Ministerial Conference in November 1971, COST is presently used by the scientific communities of 35 European countries to cooperate in common research projects supported by national funds.

The funds provided by COST – less than 1% of the total value of the projects – support the COST cooperation networks (COST Actions) through which, with EUR 30 million per year, more than 30,000 European scientists are involved in research having a total value which exceeds EUR 2 billion per year. This is the financial worth of the European added value, which COST achieves.

A 'bottom up approach' (the initiative of launching a COST Action comes from the European scientists themselves), 'à la carte participation' (only countries interested in the Action participate), 'equality of access' (participation is open also to the scientific communities of countries not belonging to the European Union) and 'flexible structure' (easy implementation and light management of the research initiatives) are the main characteristics of COST.

As precursor of advanced multidisciplinary research COST has a very important role for the realisation of the European Research Area (ERA) anticipating and complementing the activities of the Framework Programmes, constituting a 'bridge' towards the scientific communities of emerging countries, increasing the mobility of researchers across Europe and fostering the establishment of 'Networks of Excellence' in many key scientific domains such as: Biomedicine and Molecular Biosciences; Food and Agriculture; Forests, their Products and Services; Materials, Physical and Nanosciences; Chemistry and Molecular Sciences and Technologies; Earth System Science and Environmental Management; Information and Communication Technologies; Transport and Urban Development; Individuals, Societies, Cultures and Health. It covers basic and more applied research and also addresses issues of pre-normative nature or of societal importance.

Web: http://www.cost.eu

Table of contents

Part II - Proteomics and animal health

Part III - Proteomics in animal production

Proteomics in farm animals: going south

Pedro Rodrigues[1], David Eckersall[2] and André Martinho de Almeida[3]
[1]*CCMAR, Universidade do Algarve, Faro, Portugal; pmrodrig@ualg.pt*
[2]*University of Glasgow, Glasgow, Scotland, United Kingdom; david.eckersall@glasgow.ac.uk*
[3]*Instituto de Investigação Científica Tropical, Lisboa, Portugal, aalmeida@fmv.utl.pt*

Three years have passed since the Amsterdam meeting in which the founding stones of the COST Action application were first laid down. During this period, several milestone events have occurred, from a successful application procedure to the implementation of the Action initiatives. In fact, a year and a half has passed since November 2010, when COST action FA1002 – Farm Animal Proteomics was initiated in the Brussels kick off meeting. From there, FA1002 has been involved in a number of relevant initiatives, aiming to fulfill its proposed mission 'to contribute to the wide dissemination and use of Proteomics tools in Farm, Veterinary and Food Sciences'. In the subsequent lines we will address major aspects of such initiatives.

In March/April 2011, the first spring meeting took place in Glasgow, Scotland. It was an intense meeting with over 60 presentations most of them as oral communications. It was a decisive meeting for the action, which went far beyond the normal objectives of a scientific meeting: to present and discuss scientific work and achievements. In fact, the Glasgow meeting was very important because it was an opportunity for all participants in the Action to meet and to learn about each other's scientific goals, achievements and capabilities. Additionally, it was an extraordinary meeting for planning future Action initiatives and to define future collaborations between members of the Action: Short Term Scientific Missions (STSMs) and joint collaborative work and publications, not to mention the organization of a Proteomics Training School in Belvaux, Luxembourg and the decision to edit a Special Issue of Journal of Proteomics (Elsevier).

The Farm Animal Proteomics Training School took place in Luxembourg in November 2011. It was outstandingly organized locally by Jenny Renaut at the Gabriel Lippmann Institute in Luxembourg. The course had the participation of 12 young PhD students and Post-Docs from several countries of the Action: Portugal, Spain, Italy, Ireland, UK, Sweden, Norway and Slovenia; and had Trainers from Portugal, Spain, Ireland, Belgium, Italy, and Switzerland besides, of course, Luxembourg. It was a course in which students had the opportunity to learn in-depth knowledge of protein identification using mass spectrometry, particularly MALDI, among other instruments. This was a course of significant importance for students already working with proteomics and we believe it will be very important for the projects of the PhD students and young Post-Docs that attended. Nevertheless, it is our opinion that a more basic course is also needed for students starting their proteomics research and that have limited access to proteomics facilities or equipment. Such a course is planned for the Summer 2012 in Oporto, Portugal.

STSMs are a key component of any COST action. In the case of FA1002, they are particularly important because they allow the interchange of experiences and the conduction of joint and fruitful scientific collaborations. In our specific case, they allow students and researchers from institutions and countries with less prominent proteomics backgrounds and means to have access to state of the art facilities and equipment. Such an important aspect will be particularly highlighted during this forthcoming Vilamoura meeting in the form of oral communications. A similar strategy will be followed in the next round of applications for STSMs in 2012, following again a South-North and East-West rationale.

Upon the suggestion of Ingrid Miller (Vienna, Austria), FA1002 was invited to edit a special issue of Journal of Proteomics (Elsevier) on Farm Animal Proteomics. The volume will be edited by Ingrid Miller, André de Almeida and David Eckersall and the submission process of proposed contributions is currently being finalised. Submitted papers include a vast array of topics, both experimental and review papers and encompassing the large variety and potential applications of proteomics in animal and veterinary sciences. Again, scientific interactions made possible by the COST Action have allowed a significant number of contributions that include manuscripts by joint authors from two or more COST countries.

All the above initiatives bear witness to the importance of proteomics in the development of successful animal, veterinary and food sciences, as well as what can be accomplished by a determined and relentless group of scientists working in this field.

A year has passed since the Glasgow meeting. Following the underlying principle of rotating meeting organization within COST Action FA1002 and the decisions made by the MC in Glasgow, the next spring meeting will now go to the southwestern tip of Europe, to the Algarve, in Portugal. From an agriculture point of view, the Algarve is reputed for its orange and almond orchards, vineyards, as well as sheep and goat small scale farming. More recently it has become an important aquaculture production region, including farmed fish and mussels. Nevertheless, the Algarve is a region best known for a (very) mild climate, pristine beaches and historical villages and towns, being the most important tourism region of Portugal. Historically, it was the last region to be included in the kingdom of Portugal in 1249 and owes its name to its original Arab name *Al-Gharb*, meaning 'the west'. Since then, the Algarve and its inhabitants are known for a truly unique special sense of identity and charisma.

The stage is therefore set to the Vilamoura 2012 meeting. On behalf of the organizing committee we wish you all a fruitful scientific meeting. We are absolutely sure you will have a pleasant stay in the Algarve. We also would like to thank COST for its support of this publication.

Pedro Rodrigues
David Eckersall
André de Almeida

A Panorama of the Vilamoura marina (photo kindly ceded by Patricio Miguel).

Part I
Invited plenary
communications

Mass spectrometry for the veterinary and farm animal world

Renata Soares[1], Catarina Franco[1], Elisabete Pires[1], Miguel Ventosa[1,2], Rui Palhinhas[1], Kamila Koci[1], André Martinho de Almeida[1,3] and Ana Varela Coelho[1]
[1]*ITQB-UNL, Av. da República, Estação Agronómica Nacional, 2780-157 Oeiras, Portugal*
[2]*IBET, Apartado 12, 2780-901 Oeiras, Portugal; varela@itqb.unl.pt*
[3]*IICT/CVZ – Instituto de Investigação Científica Tropical, Centro de Veterinária e Zootecnia & CIISA – Centro Interdisciplinar de Investigação em Sanidade Animal, Fac. Med. Veterinária, Av. Univ. Técnica, 1300-477 Lisboa, Portugal*

Proteomics approaches are gaining increasing importance in the context of all fields of animal and veterinary sciences, including physiology, productive characterization, and disease/parasite tolerance, among others. Proteomic studies mainly aim at the proteome characterization of a certain organ, tissue, cell type or organism, either in a specific condition or comparing protein differential expression within two or more selected situations. Due to the high complexity of samples, usually total protein extracts, proteomic studies rely heavily on protein identification and quantification using mass spectrometry (MS) based methodologies or with coupling to other methodologies, like liquid chromatography and gel electrophoresis. Despite the increasing importance of MS in the context of animal and veterinary sciences studies, the usefulness of such tools is still poorly perceived by the Animal Science community. This is primarily due to limited knowledge on Mass Spectrometry by animal scientists, which use nowadays still requires a high level of specialization. Additionally, confidence and success in protein identification is hindered by the lack of information in public databases for most of farm animal species and their pathogens, with the exception of cattle (*Bos taurus*), pig (*Sus scrofa*) and chicken (*Gallus gallus*). During this lecture, a description of the main MS methodologies available for proteome characterization and differential proteomic studies will be presented.

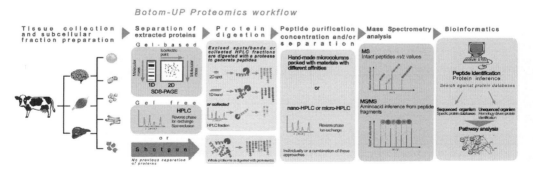

Figure 1. Schematic representation of the bottom-up proteomic workflow.

The advantages and limitations of several proteomic workflows will be discussed, together with the aspects that still need to be improved when applying them in the field of animal sciences.

Acknowledgements

The authors acknowledge the financial support from Fundação para a Ciência e Tecnologia (FCT, Lisbon, Portugal) on behalf of the National Re-equipment Program – National Network Mass Spectrometry (REDE/1504/REM/2005). Authors André M. Almeida and Renata Soares acknowledge financial support from the Ciência 2007 and 2008 programs, Catarina Franco from PhD grant SFRH/BD/29799/2006 and Miguel Ventosa a grant from project PTDC/CVT/114118/2009, all from FCT.

References

1. Chait BT. Mass spectrometry in the postgenomic era. Annual Review of Biochemistry 2011;80:239-46.
2. Sabidó E, Selevsek N, Aebersold R. Mass spectrometry-based proteomics for systems biology. Current Opinion in Biotechnology 2011.
3. Donato P, Cacciola F, Mondello L, Dugo P. Comprehensive chromatographic separations in proteomics. Journal of Chromatography. A 2011;1218:8777-90.
4. Sá-Correia I, Teixeira MC. 2D electrophoresis-based expression proteomics: a microbiologist's perspective. Expert Review of Proteomics 2010;7:943-53.

Protein synthesis pipelines for study of protein-protein interactions

Mangesh Bhide

Laboratory of Biomedical Microbiology and Immunology, Department of microbiology and immunology, University of Veterinary Medicine and Pharmacy, Kosice, Slovakia and Institute of Neuroimmunology, Slovak Academy of Sciences, Bratislava, Slovakia; mangeshbhide@hotmail.com

Discovery of the novel protein-protein interactions is a dream of many biomedical scientists. Unfolding the underlying molecular principles of biological processes like inter and intracellular events, host-pathogen interactions, cell signaling, etc. needs precise benchmarking and experimental evidence of protein-protein interactions. Over the last two decades recombinant proteins are used throughout biomedical sciences and have become an inevitable tool in the study of protein-protein interactions. Their production was once the task of experts, however the development of commercially available systems has made the technology simpler. Yet, researchers face many issues and questions like 'which is the most suitable host (bacteria, yeast, insect cells, human cells, plants or protozoa) and vector for expression?; should the protein be tagged and which affinity tag is the best?; what is a good protein purification strategy?; should one express the full-length protein or a fragment thereof?; should be protein be overexpressed in secreted or intracellular form?; should purified protein be in native or denatured state?; and so on...' These are some of the primary questions from the plethora of the difficulties that are encountered in the production of recombinant proteins. Unfortunately, because every protein is different, there can be no right answer to any of these questions.

By the 1970s, researchers had developed the ability to isolate genes or a segment of DNA that contains enough information to make one protein. By the 1980s, scientists were able to move genes from one organism to another. The first commercial application of recombinant DNA technology was used in 1982, when researchers produced human insulin for the treatment of diabetes. In the last decade scientists have changed the face of protein production, e.g. from macro scale to nano scale, *in vivo* to *in vitro* synthesis, time consuming (several days to months) to rapid (1-2 day) protocols and from classical species dependent promoter system to designing of single universal promoter for most of the expression systems.

To this background, rapid protein synthesis workflows for protein-protein interactions studies have been standardized and developed in various laboratories. These strategies include novel and rapid methods in (1) cloning like ligation independent cloning or double overlap extension based cloning, (2) use of vivid fluorescent tags, (3) rapid on-line and off-line protein purification methods, (4) use of cost effective prokaryotic and eukaryotic expression hosts like *E. coli* and *Leishmania in vivo* systems, etc. These points will be presented, step-by-step, in the first half of the presentation.

Another pipeline of recombinant protein production '*in vitro on chip technique*' is one of the most robust, rapid techniques used so far in the protein-protein interaction assays. The second part of the presentation will be dedicated to this technique.

Acknowledgements

Thanks are due to R. Mucha, S. Hresko, L. Pulzova, M. Madar, E. Bencurova, P. Mlynarcik and M. Cepkova for their immense help in experimental setup of some of the workflows presented in this plenary lecture. Financial support to setup these pipelines was from APVV-0036-10 and VEGA-2/0121/11.

Dinosaurs and other farm animals

Jane Thomas-Oates

Centre of Excellence in Mass Spectrometry and Department of Chemistry, University of York, Heslington, York, YO10 5DD, United Kingdom; jane.thomas-oates@york.ac.uk

The persistence of biomolecules, including proteins, in the archaeological record allows modern protein analytical approaches to contribute concrete scientific data to a wide range of engaging and long-standing archaeological questions. The rapid development of proteomic technologies and approaches, modern mass spectrometry's excellent limits of detection and mass accuracy, and the availability of protein sequence data from ever more animal species provide outstanding opportunities for archaeological research. In turn, the outputs of that research have direct application in farm animal proteomic studies.

For archaeological applications, tailored sample handling protocols and innovative approaches to data interpretation are proving pivotal in maximizing the contribution of mass spectrometry and proteomics to such investigations. A wide variety of archeological problems are accessible to analysis by MS. For example, fossilized bone samples may contain recoverable protein (and sometimes DNA), the sequences of which can identify the organism from which the bone came. In the case of extinct organisms, they also have the potential to reveal evolutionary links to extant species. The methods developed for such demanding archaeological analyses in turn have obvious applications in the study of modern samples.

The talk will present data obtained during investigations of both ancient and modern samples, and will illustrate a range of ways in which MS has contributed to challenging questions in both archaeology and modern animal studies.

References

1. Buckley, M., S. Whitcher Kansa, S. Howard, S. Campbell, J. Thomas-Oates, M. Collins (2010) Distinguishing between archaeological sheep and goat bones using a single collagen peptide, *J. Archaeol. Sci, 37*, 13-20.
2. Buckley M, M. Collins, J. Thomas-Oates, J.C. Wilson (2009) Species identification by analysis of bone collagen using matrix-assisted laser desorption/ionisation time-of-flight mass spectrometry, *Rapid Commun. Mass Spectrom. 23*, 3843-3854.
3. Heaton, K., C. Solazzo, M. Collins, J. Thomas-Oates, E. Bergström (2009) Towards the application of desorption electrospray ionisation mass spectrometry (DESI-MS) to the analysis of ancient proteins from artefacts. *J Archaeol. Sci. 36*, 2145-2154.
4. Buckley, M., M.J. Collins, J. Thomas-Oates (2008) A Method of Isolating the Collagen (I) α2 Chain Carboxytelopeptide for Species Identification in Bone Fragments. *Anal. Biochem. 374*, 325-334.
5. Asara, J.M., M.H. Schweitzer, L.M. Freimark, M. Phillips, L.C. Cantley (2007) Protein Sequences from Mastodon and *Tyrannosaurus rex* Revealed by Mass Spectrometry. *Science 316 (5822)*, 280-285.

6. Buckley, M., A. Walker, S.Y.W. Ho, Y. Yang, C. Smith, P. Ashton, J. Thomas Oates, E. Cappellini, H. Koon, K. Penkman, B. Elsworth, D. Ashford, C. Solazzo, P. Andrews, J. Strahler, B. Shapiro, P. Ostrom, H. Gandhi, W. Miller, B. Raney, M. I. Zylber, M.T.P. Gilbert, R.V. Prigodich, M. Ryan, K.F. Rijsdijk, A. Janoo, M.J. Collins (2008) Comment on "Protein Sequences from Mastodon and *Tyrannosaurus rex* Revealed by Mass Spectrometry", *Science, 319*, 33 (2008); DOI: 10.1126/science.1147046.

Proteomics strategies to trace illegal growth-promoters in cattle

A. Urbani[1;2], C. Nebbia[3], M. Carletti[3], G. Gardini[3], D. Bertarelli[3], M. Ronci[4], L. Della Donna[3] and P. Sacchetta[5]
[1]University of Rome 'Tor Vergata', Rome, Italy; andrea.urbani@uniroma2.it
[2]IRCCS-Fondazione Santa Lucia Centro di Ricerca sul Cervello, Rome, Italy
[3]Department of Animal Pathology, Division of Pharmacology and Toxicology, Faculty of Veterinary Medicine, Università di Torino, Grugliasco (Turin), Italy
[4]Centro Studi sull'Invecchiamento (Ce.S.I.), Chieti, Italy
[5]Dipartimento di Scienze Biomediche, Università 'G. D'Annunzio' di Chieti-Pescara, Italy

Despite the intensive control of illegal growth promoting agents such as sexual steroids, corticosteroids and β-agonists carried out within the European Union (EU) in cattle and other food producing species, over the last few years the number of reported positives has been very low, averaging about 0.2-0.3%. There is evidence, however, that these figures may underestimate the real incidence of GP abuse in meat cattle breeding. Besides the risk for animal health, residual amounts of these chemicals in the resulting animal products are dangerous for the consumers. The low rate of reported positives would suggest the use of a combination of different active principles either at very low dosages and of unknown chemical structure not included in the national residue monitoring plans.

There is therefore the urgent need to develop biological assays that could ascertain the exposure of food producing animals to a number of drugs or hormones currently used for the chemical manipulation of animal growth. The introduction of novel strategies based on the biological functional impact of these compound classes may eventually provide a reliable and cost effective detection method.

We have developed several technological tools in order to collect and compare large multivariate dataset on protein repertoire. Mass spectrometry could represent an innovative and useful approach to provide the extremely accurate data highlighting the eventual modifications of protein expression resulting from the exposure of living animals to such agents. We have focused toward the development of a robust proteomics platform based on mass spectrometry to investigate the protein repertoire of bovine sera and tissue with the double purpose to characterize the protein profile and to possibly highlight molecular features of illicit treatments comparing treated and non treated animals.

Through MALDI-TOF/TOF-MS and shotgun UPLC-ESI-QTOF-MS/MS ion accounting acquisition strategy, we have analyzed Friesian male calves treated with different GPAs such as: 17-β-estradiol, clenbuterol and dexamethasone.

Our efforts have been initially focused on the low range proteome repertoire of sera in the peptide and proteins range detectable between 3-20 kDa. This specific range represents a serum fraction of great biological relevance since it encodes information on protein and peptides, proteolytic maturation and post-translational modifications. The high analytical performances achieved with a MALDI-TOF-MS platform in this mass window allow a fine micro-characterisation of molecular species. The study of the low molecular range was also suggested by the intent to investigate proteins in a range not easily detectable from other common proteomic techniques, such as 2D electrophoresis and SDS-PAGE. Our results concerning inter- and intra-day repeatability confirm CV% lower than the limits for full analytical validation imposed by several international organizations, including the FDA. The high mass resolution (FWHM >1000) and accuracy (~100 ppm) allows the analysis to clearly identify specific protein products characteristic of calf sera. The overall evaluation of these analytical features plainly demonstrates how a fine tuned proteomic platform, based on MALDI-TOF-MS, can be employed for semi-quantitative molecular biomarker investigations. Moreover this platform can be directly extended to the morphological mapping of discriminant signals in target organs based on Imaging Mass Spectrometry.

Molecular imaging using MALDI-TOF mass spectrometry (MALDI-IMS) is a recently developed technology which joins the protein characterization properties of the matrix desorption ionization via laser pulses with the possibility to directly analyse thin tissue sections, thus enabling the correlation of the molecular profile with the tissue distribution. The application of such investigations was initially developed on fresh or cryo preserved tissue specimen. More recently we have improved the unlocking procedure thus opening the route to the employment of FFPE tissue samples in the molecular definition of potential multifactorial biomarkers. Moreover the possibility to release peptides from FFPE samples, amenable to proteomics analysis, has pushed these protocols toward their application in shotgun proteomics experiments via high sensitivity LC-MS^E.

Samples analyzed through a shotgun proteomics approach by UPLC-ESI-QTOF-MS^E returned the identification and the quantification relative to reference standard without stable isotope labeling. The golden standard for the quantification of proteins on hyphenated instruments quadrupole/time-of-flight is 'data dependent acquisition' (DDA): the quadrupole works in transmission mode during all the run and only when the ion intensity of a certain specie reaches the threshold the quadrupole switches to MS/MS mode acquiring the fragments of that ion. The specificity for this kind of acquisition is very high, nonetheless, when dealing with complex mixtures, co-elution phenomena of different species commonly happens and the less abundant are not recorded. Parallel acquisition also called 'data independent' MS^E is able to bypass this problem, with the continuous and rapid switching between MS and MS/MS mode during all the chromatographic run. Finally two MS tracks are recorded: one at low collision energy and the other at high collision energy; the tracks are then aligned and merged together to have the list of all parents with the corresponding fragments. Such a type of acquisition let very complex mixtures to be effectively analyzed.

Farm animal proteomics

Specific disciminant molecular signatures have been associated with the GPAs treatment; the overall collected evidences will be introduced and discussed during the presentation.

References

1. Nebbia C, Urbani A, Carletti M, Gardini G, Balbo A, Bertarelli D, Girolami F. Novel strategies for tracing the exposure of meat cattle to illegal growth-promoters. Vet J. 2011 Jul;189(1):34-42. Epub 2010 Jul 24. Review. PubMed PMID: 20659808.
2. Della Donna L, Ronci M, Sacchetta P, Di Ilio C, Biolatti B, Federici G, Nebbia C, Urbani A. A food safety control low mass-range proteomics platform for the detection of illicit treatments in veal calves by MALDI-TOF-MS serum profiling. Biotechnol J. 2009 Nov;4(11):1596-609. PubMed PMID: 19844911.
3. Gardini G, Del Boccio P, Colombatto S, Testore G, Corpillo D, Di Ilio C, Urbani A, Nebbia C. Proteomic investigation in the detection of the illicit treatment of calves with growth-promoting agents. Proteomics. 2006 May;6(9):2813-22. PubMed PMID: 16572471.
4. Maddalo G, Petrucci F, Iezzi M, Pannellini T, Del Boccio P, Ciavardelli D, Biroccio A, Forlì F, Di Ilio C, Ballone E, Urbani A, Federici G. Analytical assessment of MALDI-TOF Imaging Mass Spectrometry on thin histological samples.
5. An insight in proteome investigation. Clin Chim Acta. 2005 Jul 24;357(2):210-8. PubMed PMID: 15913587.

Assessing fish quality in aquaculture: a proteomics approach

Pedro M. Rodrigues, Nadége Richard, Mahaut de Vareilles, Tomé S. Silva, Odete Cordeiro, Luis E.C. Conceição and Jorge Dias
CCMAR, Centro de Ciências do Mar do Algarve, universidade do Algarve, Campus de Gambelas, 8005-139 Faro, Portugal; pmrodrig@ualg.pt

Aquaculture has shown an amazing growth rate in fish production over the last decade, though nowadays it is a very competitive market in the food industry. Fish and shellfish demand will continue to grow, in part as a response to population growth. Provision of seafood from capture fish is declining and is partly not sustainable. Seafood from aquaculture will potentially overcome this supply issue. It can deliver a product of defined quality, composition and safety to the market in all seasons of the year enabling a greater penetration of fish products in consumer's diet.

Fish quality is a broad and complex concept. However, from a general standpoint major quality attributes in fish are: freshness, safety, nutritional and health value, eating quality, appearance and authenticity. Flesh quality in fish is species-dependent, results from a complex set of intrinsic traits such as the muscle chemical composition (fat content and fatty acid profile, glycogen stores, oxidative stability, colour) and muscle cellularity and is strongly influenced by a variety of extrinsic factors such as feeding, pre- and post-slaughter handling, processing and storage procedures. In brief, all factors affecting flesh quality can be divided into two aspects on the genetic basis and on management systems.

An awareness regarding the use of scientific knowledge and emerging technologies to obtain a better farmed organism through a sustainable production has enhanced the importance of proteomics within this industry. During the last decade Omic technologies (i.e. genomics, metabolomics and proteomics) have been widely implemented in the field of farm animal proteomics with a very positive impact in areas such as aquaculture. Proteomics in particular has emerged as a powerful tool towards a deep understanding of marine organisms' biology, helping aquaculture to reach its main goal; a better quality product. The major drawback in aquaculture proteomics is the lack of information at the genome level in most of the 310 cultured species reported by FAO in 2008. As a result, only a limited number of fully sequenced proteins is available.

Farmed seafood organisms are susceptible to a wide range of factors that can pose a major threat to a thriving aquaculture industry with considerable economical repercussions. This industry has been going through major challenges in its effort to respond to the continuous higher consumer demand, coupled with clear market global awareness of a better quality product and also animal welfare. A good balance between these challenges may greatly benefit from a better scientific understanding of the biological traits in seafood farming.

We approach this issue looking at factors in fish farming like welfare, nutrition and health, by integrating comparative two-dimensional proteome analysis in the production fish process as can be seen in Figure 1.

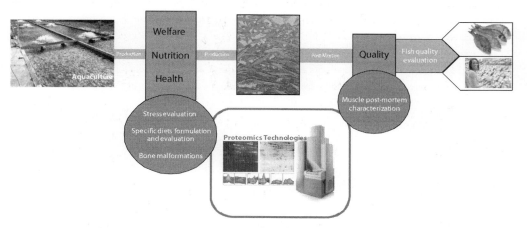

Figure 1. Proteomics approach for quality assessment and evaluation in fish aquaculture.

In this lecture a brief description regarding the proteome characterization and methodologies used by our group to achieve a better quality cultured fish will be presented. An overview of possible biomarkers will also be presented.

References

1. Pedro M. Rodrigues, Tomé Silva, Jorge Dias, Flemming Jessen (2012). Proteomics in aquaculture: applications and trends. *Journal of Proteomics.* Submitted.
2. Tomé S. Silva, Odete D. Cordeiro, Elisabete D. Matos, Tune Wulff, Jorge P. Dias, Flemming Jessen, Pedro M. Rodrigues (2012). Effects of pre-slaughter stress levels on the postmortem sarcoplasmic proteomic profile of gilthead seabream muscle. *Journal of Agricultural and Food Chemistry.* Submitted.
3. Mahaut de Vareilles, Nadège Richard, Paulo J. Gavaia, Tomé S. Silva, Odete Cordeiro, Inês Guerreiro, Manuel Yúfera, Irineu Batista, Carla Pires, Pedro Pousão-Ferreira, Pedro M. Rodrigues, Ivar Rønnestad, Kari E. Fladmark, Luis E.C. Conceição(2012) Impact of dietary protein hydrolysates on skeleton quality and proteome in *Diplodus sargus* larvae. *Journal of Applied Ichthyology.* In Press.

4. Odete D. Cordeiro, Tomé S. Silva, Ricardo N. Alves, Benjamin Costas, Tune Wulff, Nadège Richard, Mahaut de Vareilles, Luís E.C. Conceição, Pedro M. Rodrigues (2012). Chronic stress conditions and liver proteome expression in Senegalese sole (*Solea senegalensis*): identification of potential welfare indicators. *Marine Biotechnology*. In press.

5. Tomé S. Silva, Odete Cordeiro, Nadège Richard, Luis E.C. Conceição, Pedro M. Rodrigues (2011) Changes in the soluble bone proteome of reared white sea bream (*Diplodus sargus*) with skeletal deformities. *Comparative Biochemistry and Physiology, Special Issue Part D: Genomics and Proteomics.* D 6, 82-91.

6. Tomé S Silva, Odete Cordeiro, Flemming Jessen, Jorge Dias, Pedro M Rodrigues (2010). On the reproducibility of a fractionation procedure for fish muscle proteomics. *American Biotechnology laboratory.* Proteomics special issue May/June 2010, 8-13.

7. Ricardo N. Alves, Odete Cordeiro, Tomé S. Silva, Nadège Richard, Mahaut de Vareilles, Giovanna Marino, Patrizia Di Marco, Pedro M. Rodrigues and Luís E.C. Conceição (2010). Metabolic molecular indicators of chronic stress in gilthead seabream (*Sparus aurata*) using comparative proteomics. *Aquaculture,* 299 (1-4), 57-66.

From meat to food: the proteomics assessment

Andrea Mozzarelli[1,2,3], Gianluca Paredi[1,2], Barbara Pioselli[1] and Samanta Raboni[1,2]
[1]Department of Biochemistry and Molecular Biology, University of Parma, Italy;
andrea.mozzarelli@unipr.it
[2]Interdepartmental Center Siteia.Parma, University of Parma, Parma, Italy
[3]National Institute of Biostructures and Biosystems, Rome, Italy, Parma, Italy

Farm animal meat is a worldwide food with poultry, pork and bovine meat supplying a large portion of proteins to human diet. Because animal muscles cannot be consumed immediately after slaughtering, aging is required to achieve the transformation of muscles to meat. This process is associated with a complex pattern of molecular events that depends on many variables, including animal genetic profile, diet, slaughtering procedures and temperatures. Correlations between such variables and protein pattern have been and are still being thoroughly studied applying proteomic methods [1-4].

In some countries, such as Spain, France, Italy and Portugal, pork meat is very frequently and intensively used for manufacturing food products aimed at lasting at room temperature for days, weeks or even years. The pork food products include cooked and dry cured ham, smoked bacon, black pudding and fresh or smoked pork sausages. Whereas tenderness is the main quality for bovine meat, the most important properties for pig meat are taste, color and water holding capacity, that, in turn, are associated with consistence and palatability. Salt solutions (brine) or salt crystals are used in pork meat treatment to prevent degeneration and to confer stability and consumer-attractive color and taste.

We will discuss some of the proteomic studies that have been carried out on cooked ham and dry cured ham carried out with the aim of correlating meat processing with quality. Ham manufacturing is still dominated by traditional recipes and the scientific contributions are limited to the determination of a few properties, such as pH, conductivity, bacterial load, color degree, tenderness.

Cooked ham. The cooked ham production involves three main steps:
1. Injection into meat pieces of brine at different salt concentration, usually between 15 and 45%, containing other ingredients, such as casein or spices.
2. Tumbling of meat for different time lengths, at selected temperatures. This step leads to extensive muscle cell breakage with formation of a protein exudate that covers the muscle pieces and acts as a glue between them.
3. Cooking of the assembled meat pieces at 68-72 °C to reduce the microbial load. Ham is then refrigerated, packaged and aged for at least one month.

Proteomic methods were exploited to characterize the exudate protein pattern as a function of processing conditions [5]. The effect of different brine concentrations from 15 to 45% and tumbling process conditions (4-10 °C for 16-24 hours) on the protein profile of the exudates were determined by 2D-GE and mass spectrometry. The nine most abundant spots, that are associated with six proteins and accounted up to the 90% of protein quantity, were identified by peptide mass fingerprinting. Myosin light chain 1 and myosin regulatory light chain were found in higher concentrations in exudates at 15 and 30%, whereas tropomyosin alpha chain was more abundant in exudate at 30%. Concentration of tropomyosin beta chain was two fold higher in sample obtained after 16 hours of tumbling at 10 °C with respect to samples tumbled at 4 °C. Three spots, identified as actin alpha skeletal muscle, vary in intensity according to the different tumbling conditions. Results showed that there are correlations between each protein profile and processing conditions. This in turn suggests that the different protein profiles affect the ham texture, and, hence, the quality of the final product. Protein profiles thus may qualify as quality biomarker during cooked ham processing.

Dry cured ham. The dry cured ham is obtained through a process that consists of three phases: (1) salting; (2) resting; and (3) drying and maturation:
1. Salting of pork leg pieces is carried out by covering it with both dry and wet salt and storing the meat for two-three weeks at 1-3 °C with relative humidity higher than 80%. Salt slowly diffuses inside the meat, breaking muscle cells and producing a protein exudate.
2. Salted pork meat rests in an air circulating environment, at 2-4 °C and low humidity, causing meat dehydration.
3. Drying and maturation steps last at least one year, with the ham stored under air circulation at 14-18 °C and low humidity.

These steps vary from one producer to another as well as from one country to another according to traditional recipes that are influenced by pork meat quality and environmental conditions, i.e. temperature and air humidity. The resulting meat dehydration, protein solubilisation and proteolysis, lipolysis and formation of volatile molecules generate sensory features that are typical of each brand of dry cured ham.

Proteomic methods have been exploited in the characterization of each step. Two main muscles are present in cured ham, the *semimembranosus* and *biceps femoris*, being located on the surface and in the internal part, respectively. The proteomic investigations have determined that proteins related to energy metabolism were overpopulated in the *semimembranosus*, whereas myofibrillar proteins are more abundant in *biceps femoris.* This observation can be explained by a lower penetration of salt in *biceps femoris.* Furthermore, different levels of salt lead to a different degree of proteolysis, explaining the high number of protein fragments identified in the *biceps femoris* with respect to *semimembranosus* [6]. The effect of salt concentration on proteolysis was also investigated during the ripening phase. Both the myofibrillar and sarcoplasmic proteins undergo major changes as shown by 2D-GE map of water soluble protein fraction of the raw meat and the final product obtained upon 14 months of ripening. Changes were observed for creatine kinase, enolase B, glyceraldehyde 3-phosphate dehydrogenase, tropomyosin alpha and beta chain expression. Some spots appeared and later disappeared indicating fragments that are formed and further processed [7-10].

The differential effect of the level of salt on the *biceps femoris* from PRKAG3 and CAST pig genotypes was investigated with 2D-GE. Results showed that the PRKAG3 genotype is mainly associated with proteins related to metabolism, enolase 3, muscle creatine kinase, glycerol-3-phosphate dehydrogenase, albumin and annexin. Within the two CAST genotypes 13 spots showed different concentrations. Three exhibited a higher concentration in CAST1 genotype, while the other ten exhibited a higher concentration in CAST2 genotype. This investigation also showed that the degree of salt strongly affects the proteolytic activity. Furthermore, the abnormal glycogen contents of PRKAG3 pig muscle affects the post-mortem metabolic potentials and muscle pH decline [11].

Given the higher commercial value of processed with respect to untreated pig meat, the number of so far published proteomic investigations is surprisingly low. Further studies are needed to fully characterize the industrial transformation for both dry cured and cooked ham, thus contributing to processing optimization. Furthermore, as salt is an increasing concern for health, a full understanding of the events that occur during the processing conditions may help to lower salt concentrations during the transformation process. The challenge is to produce a high quality ham at a lower salt content.

References

1. Bendixen E, Danielsen M, Hollung K, Gianazza E, Miller I. Farm animal proteomics – A review. J Proteomics. 2010;74:282-93.
2. Hollung K, Veiseth E, Jia X, Faergestad EM, Hildrum KI. Application of proteomics to understand the molecular mechanisms behind meat quality. Meat Sci. 2007;77:97-104.
3. Bendixen E. The use of proteomics in meat science. Meat Sci. 2005;71:138-49.
4. Paredi G., Raboni S., Bendixen E., Almeida AM, Mozzarelli A. 'Muscle to meat' molecular events and technological transformations: the proteomics insight. J Proteomics. *Submitted*
5. Pioselli B, Paredi G, Mozzarelli A. Proteomic analysis of pork meat in the production of cooked ham. Mol Biosyst. 2011;7:2252-60.
6. Théron L, Sayd T, Pinguet J, Chambon C, Robert N, Santé-Lhoutellier V. Proteomic analysis of *semimembranosus* and *biceps femoris* muscles from Bayonne dry-cured ham. Meat Sci. 2011;88:82-90.
7. Sentandreu MÁ, Armenteros M, Calvete JJ, Ouali A, Aristoy M-C, Toldrá F. Proteomic identification of actin-derived oligopeptides in dry-cured ham. J Agric Food Chem. 2007;55:3613-9.
8. Mora L, Sentandreu MA, Toldrá F. Identification of small troponin T peptides generated in dry-cured ham. Food Chem. 2010;123:691-7.
9. Sforza S, Boni M, Ruozi R, Virgili R, Marchelli R. Identification and significance of the N-terminal part of swine pyruvate kinase in aged Parma hams. Meat Sci. 2003;63:57-61.

10. Okumura T, Yamada R, Nishimura T. Survey of conditioning indicators for pork loins: changes in myofibrils, proteins and peptides during postmortem conditioning of vacuum-packed pork loins for 30 days. Meat Sci. 2003;64:467-73.

11. Škrlep M, Čandek-Potokar M, Mandelc S, Javornik B, Gou P, Chambon C, *et al.* Proteomic profile of dry-cured ham relative to PRKAG3 or CAST genotype, level of salt and pastiness. Meat Sci. 2011;88:657-67.

Ruminant saliva: accessory proteins of homeostasis, olfaction and defence

Marcus Mau
Saliva Research Group, Dental Institute, Floor 17, Tower Wing, Guy's Hospital, King's College London, London, SE1 9RT, United Kingdom; marcus.mau@kcl.ac.uk

Introduction

Mostly fed with grass in fresh or conserved form, cattle and other grazing ruminants such as goat, sheep and camels have to cope with two main disadvantages of their food. On one side, silicate defence bodies from plant cells (phytoliths), and environmental grit are supposed to abrade tooth enamel (Baker *et al.*, 1959). Furthermore, it seems likely that these silicates could bind different important salivary proteins during mastication with yet indefinite effects on pH regulation, olfaction or anti-bacterial defence (Mau *et al.*, 2006). However, the interactions of salivary proteins with inorganic food particles in free-ranging and livestock animals are still unknown.

On the other side, in addition to environmental silicates, grasses and leaves contain large amounts of structural carbohydrates such as cellulose, which cannot be digested by the animals themselves. Therefore, animals, such as cattle, feeding on cellulose-rich materials depend on the digestive assistance of symbiotic, cellulolytic microorganisms that live in specialized compartments of their stomach or gut (Zilber-Rosenberg & Rosenberg, 2008). In fact, the microorganisms that digest the cellulose for their hosts are oxygen- and pH-sensitive, resulting in a demand for a regulated pH environment inside the fermentative chamber provided by salivary secretions (Kay, 1960; Phillis, 1976). Thus, the expression of specific enzymes (carbonic anhydrases), controlling the homeostasis of the digestive tract in ruminating animals could elucidate salivary adaptations to provide and/or maintain the specialized functions of the ruminant digestive system.

This conference contribution summarizes our own, partially published work concerning accessory salivary proteins in cattle that are secreted into saliva and differentially bound to dental tissues and silicates. Accessory proteins are herein defined as those proteins that might not have a primary function in food procession or digestion, but in homeostasis, olfaction and anti-bacterial defence.

Material and methods

Saliva samples

Saliva of three cows was obtained by natural flow into a beaker at the Leibniz Institute for Farm Animal Biology (FBN) in Dummerstorf, Germany. After subsequent precipitation of food debris by centrifugation at 300xg for 5 minutes, supernatants were harvested and stored at -80 °C.

Identification of salivary carbonic anhydrases

Two-dimensional gel electrophoresis and subsequent immunoblotting were performed to visualize salivary carbonic anhydrase isoenzymes in bovine saliva using two different CA-specific antibodies as it has been described in details before (Mau *et al.*, 2010). Whole saliva samples containing 100 µg of total protein were subjected to isoelectric focussing (IEF) at 20 °C on 11 cm IPG strips pH 4-7. After that, proteins on the IPG strips were equilibrated and applied on top of a 12% SDS-PAGE gel to separate the proteins for 2 h with a constant current of 30 mA (Mau *et al.*, 2010). The gel was then silver-stained and CA-protein spots earlier identified by immunoblotting were cut out and analysed by MS mass spectrometry according to the procedure described below. BSP30 served as a reference protein to check the validity of the identification method.

Phytolith and dental preparations

For phytolith extraction from grass/hay pellets a dry ash method was used according to Parr *et al.* (2001) using a muffle oven, 10% HCl and 15% H_2O_2 for washing.

For dental powders, lower bovine jaws of two cows were obtained from the FBN in Dummerstorf, Germany, and were prepared according a method earlier described by Mau *et al.* (2006). For the experiments, teeth were thawed and 4x4x2 mm discs were cut. After washing the discs in distilled water in an ultrasonic bath (Bandelin Sonorex RK 100) for 5 min, enamel and dentine were separated. Dried enamel and dentine were ground separately in a mortar and the resulting powder fractions were sieved using a mesh cascade of 300 µm, 200 µm, 180 µm, 125 µm and 90 µm. The powder <90 µm was weighed and kept in 1.5 ml reaction tubes.

Binding experiments and MS mass spectrometry

After washing, 6 mg of each powder were weighed into different 1.5 ml reaction tubes. Samples were then added either 200 µl bovine whole saliva or 200 µl of distilled water (negative controls) and incubated for 4 h at 37 °C with constant agitation.

After incubation, powder samples were washed 6 times with 200 μl 0.9% NaCl solution and centrifuged at 2,000x*g* for 5 min. Samples with remaining, bound proteins were added 10 μl reducing loading buffer, heated and run on 10% resolving gels in 1D-SDS-PAGE. Right after electrophoresis, gels were stained with Coomassie-R250 over night and after that bands of interest were cut out and analysed by MS mass spectrometry.

Protein bands bound to enamel, dentine or silicates were excised from three different replicates. Gel pieces were subsequently de-stained, and digested with trypsin. Peptides were acidified with formic acid, and loaded via auto-sampling to a C_{18} PepMap100 column (HPLC ultimate 3000; Dionex, Sunnyvale, CA, USA).

Protein identification was done on a MS HCTultra PTM Discovery system (Bruker Daltonics, Billerica, MA, USA). The generated mass spectra were used to search the NCBI protein database with the help of Data Analysis, Version 3.4.192.1 (Bruker Daltonics), and the Mowse MASCOT-software (Matrix-Science, Boston, USA). Individual thresholds of 95% ($P<0.05$) for Mowse were considered as confident protein identifications.

Results

Beside CA-VI a second carbonic anhydrase, CA-II, was present in bovine saliva and both enzymes were further identified in saliva of other ruminating animals, such as goats and camels (Mau *et al.*, 2009, 2010). Using an anti-bovine CA specific antibody (AbD Serotec) both CA-VI and CA-II reacted positive, with molecular weights of 42 kDa (CA-VI) and 29 kDa (CA-II). Using an anti-human CA-II specific antibody (Santa Cruz) only the secreted CA-II isoform was detected. The immunoreactions showed the localization of CA-VI and CA-II in the bovine salivary proteome separated by 2D gel electrophoresis (Figure 1A). CA-

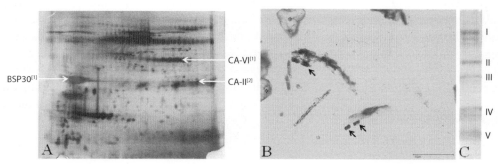

Figure 1. (A) Two-dimensional electrophoresis of bovine whole saliva and single proteins BSP30, CA-VI and CA-II. [1]: identified by mass spectrometry. [2]: identified by immunoblotting only. (modified from Mau et al., *2010). (B) Isolated silicates from plant material: phytoliths (arrows) and grit (asterisks). (C) Silicate-bound bovine salivary proteins that were further identified using MS mass spectrometry. I: lactoperoxidase; II: CA-VI; III: BSP30; IV: OBP; V: haemoglobin β.*

VI and BSP30 (reference protein) were additionally identified by mass spectrometry, whereas the CA-II refused further MS identification in the attempts.

Tooth enamel and dentine mainly bound bovine odorant-binding protein (bOBP), BSP30 and carbonic anhydrase VI (CA-VI), whereas the phytolith/silicate fraction (Figure 1B) showed additional strong interactions with haemoglobin β and lactoperoxidase (Figure 1C). Table 1 summarizes the five salivary proteins that interacted with silicates and/or dental material *in vitro* and have been identified by MS mass spectrometry.

Table 1. Identification results of silicate-binding bovine salivary proteins.

No.	Protein name	Protein accessio n no. (NCBI)	Total protein score	Molecular weight [kDa]	Sequence coverage [%]
I	Lactoperoxidase	P80025	171	80.59	13
II	Carbonic anhydrase VI (CA-VI)	P18915	175	36.98	30
III	BSP30	P79124	260	26.37	33
IV	Odorant-binding protein (OBP)	P07435	555	18.49	67
V	Haemoglobi n β	P02070	354	15.94	61

Conclusions

The differential binding of bovine salivary proteins to dental tissues and environmental silicates revealed an interesting cluster of proteins with accessory functions in pH-regulation, oxygen-binding and antimicrobial defence. Together salivary carbonic anhydrases, BSP30, lactoperoxidase and haemoglobin may provide a defence and regulatory network in ruminant saliva. This protein network is hypothetically basic to secure and protect the oxygen- and pH-sensitive microenvironment in the rumen. Thus it would guarantee successful and undisturbed cellulose fermentation, which is essential for host survival. However, more detailed work is needed on the origins, functions and dependencies of accessory proteins in bovine saliva.

Acknowledgements

The work on ruminant salivary proteins has been supported by the DFG (grant: SU124/15-1).

References

Baker G, Jones LHP, Wardrop ID. *Nature* 1959;184:1583-4.
Kay RN. J Physiol 1960;150:515-37.

Mau M, Müller C, Langbein J, Rehfeldt C, Hildebrandt JP, Kaiser TM. *Arch Anim Breed* 2006;49:439-46.

Mau M, Kaiser TM, Südekum KH. *Arch Oral Biol* 2009;54:354-60.

Mau M, Kaiser TM, Südekum KH. *Histol Histopathol* 2010;25:321-9.

Parr JF, Dolic V, Lancaster G, Boyd WE. *Rev Palaeobot Palynol* 2001;116:203-12.

Phillis J. Motility and secretions of the various regions of alimentry tract. In: J. Phillis (ed): Veterinary physiology. 1976:416-459.

Zilber-Rosenberg I, Rosenberg E. *FEMS Microbiol Rev* 2008;32:723-35.

Proteomics of foes and friends adapted to milk environment: focus on Gram-positive bovine mastitis bacteria and probiotics

Pekka Varmanen[1], Kerttu Koskenniemi[2], Pia Siljamäki[3], Tuula A. Nyman[3] and Kirsi Savijoki[1]
[1]*Department of Food and Environmental Sciences, University of Helsinki, Finland;*
pekka.varmanen@helsinki.fi
[2]*Department of Veterinary Biosciences, University of Helsinki, Finland*
[3]*Institute of Biotechnology, University of Helsinki, Finland*

The evolutionary success of bacteria depends on their excellent genetic flexibility that promotes adaptation in diverse environmental niches and different host organisms. The economically important group of bacteria well-adapted to milk-environment include species of the genera like *Staphylococcus*, *Streptococcus* and *Lactobacillus*. Some of these species are important causative agents of infectious diseases such as bovine mastitis, while many are able to form commensal relationship with the host or even able to positively affect the health. Bovine mastitis is most frequently caused by *Staphylococcus aureus*, *Staphylococcus epidermidis* and *Streptococcus uberis* strains and it is a major economic burden in the dairy industry world-wide. It is still the most common reason for antibiotic use in dairy farms, which has raised concerns on development of antimicrobial resistance and residues in the food chain. In most bacteria, the activation of SOS response has been considered the major driving force in increasing genomic fitness and adaptation (via mutagenesis, acquisition pathogenic islands, prophages or drug resistance genes, etc.).

An intriguing distinction between *Staphylococcus*, *Streptococcus* and *Lactobacillus* species is that staphylococci and lactobacilli seem to employ the classical LexA-regulated SOS response to improve viability, whereas streptococci are devoid of the classical LexA-regulated SOS response. Instead, *Streptococcus* species appear to employ varying mutagenesis and genomic variation promoting mechanisms, such as genetic transformability or HdiR-regulated SOS-response [1-3].

Current consensus is that all antibiotics will eventually be fought off by bacteria, which gives rise to new even more virulent strains. This highlights the need for next-generation treatment strategies with novel modes of action to combat ongoing evolution of resistance and the development of highly virulent super-bugs. To be able to interfere with the adaptation mechanisms, an unbiased view of changes occurring in metabolism of bacteria that most frequently are exposed to antibiotics is needed. This can only be reached by multiproteome screen combined with expression analyses of pathogenic and commensal bacteria of different origin to pinpoint central regulatory mechanisms boosting adaptation (Figure 1).

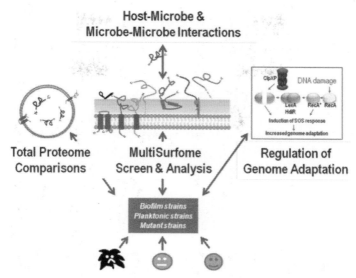

Figure 1. Multigenome and proteome screen and expression analyses of both commensal (in green & orange) and pathogenic bacteria (in black) are needed to combat pathogens with significant genome plasticity.

In our ongoing research we use comparative genomics and proteomics, analyses of bacterial responses to (antibiotic) stress conditions and exposure to host components to meet these goals.

Our comparative approaches have, for example, included multigenome screen and analysis of several *Staphylococcus* species and strains to indicated host specificity determining factors encoded by a bovine mastitis causing *Staphylococcus epidermidis* strain (mSP). The data gathered from these comparisons (Figure 2) were complemented with both quantative and qualitative proteomics [4-6]. These proteome-level analyses have included the mSP strain, as well as, infectious, biofilm forming RP62A strain and commensal human strain ATCC12228 with low infection potential. Using similar approaches we have investigated different *Lactobacillus rhamnosus* and *Lactobacillus gasseri* strains isolated from different environments (human, dairy products) to search for mechanisms contributing to adaptation in the host [7-9] (Figure 4). To expand our knowledge in molecular mechanisms modulating probiotic functions, we have complemented the proteomics studies with transcriptome analyses to thoroughly assess the effects of some relevant stress factors [10-12]. The adaptation mechanisms of bovine mastitis bacteria to antibiotic (fluoroquinolone) stress has been studied with *S. uberis* as a model [13] and the effect of penicillin exposure on protein expression has recently been investigated in *S. aureus* biofilms (Figure 5). It is well-established that biofilm growth dramatically increases resistance to antibiotics. Both studies clearly demonstrated that these distinct species adjust their metabolism to favour increasing mutagenesis and adaptation. In case of *S. aureus*, this was demonstrated by decreased DNA repair activity that

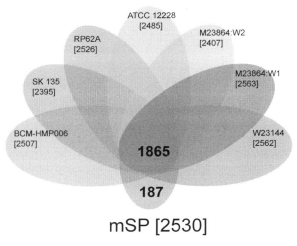

Figure 2. Venn diagram illustrating multigenome screen of Staphylococcus epidermidis *genomes to screen for conserved and unique genes in a mastitis causing* S. epidermidis *strain [3].*

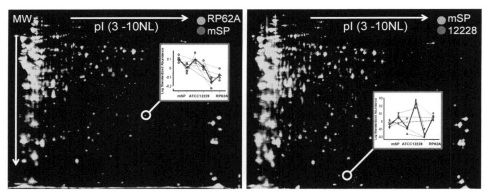

Figure 3. 2D DIGE images showing the comparison of three different Staphylococcus epidermidis *strains of bovine and human origin [4].*

was accompanied with increased synthesis of virulence factors [unpublished]. We recently demonstrated that the mastitis causing bacterium *S. uberis* is also capable of biofilm growth and that specific protein components of bovine milk strongly stimulate biofilm formation [14]. *S. uberis* strains often cause persistent infections, even though isolates are mainly susceptible to antibiotics. Thus, ability for biofilm growth could provide one explanation for persistence of *S. uberis* infections and reveal its capability for adaptation. Previously, we have demonstrated the potential of this bacterium for adaptive mutagenesis and have shown that *S. uberis* employs distinct mechanisms for fluoro-quinolone-/UV-induced mutagenesis, which is a striking difference to *Escherichia coli* SOS model [15].

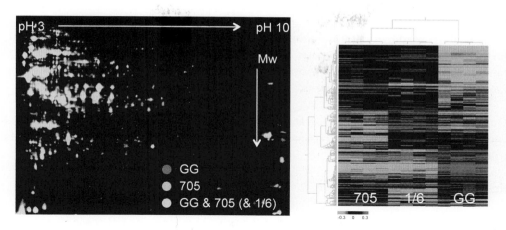

Figure 4. 2D DIGE image, dendrogram and heat map of differentially expressed proteins from three Lactobacillus rhamnosus *strains with average volume ratios ≥1.5-fold (P<0.01) and clustering of the individal spot aps [Koskenniemi, unpublished].*

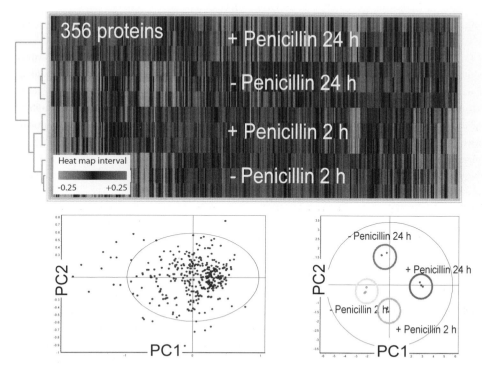

Figure 5. Dendrogram, heat map and PCA plot of differentially expressed proteins from Staphylococcus aureus *biofilms with or without a 2-hour and 24-hour treatment with penicillin at 1 mg/m.*

Proteome analysis of *S. uberis* under antibiotic exposure suggested that imbalanced dNTP pools could play role in fluoroquinolone induced mutagenesis in this bacterium [13]. To further explore the new mutagenesis mechanism in *S. uberis* we have e.g. constructed a random mutagenesis library of *S. uberis* Δ*umuC* strain in order to identify genes affecting fuoroquinolone induced mutagenesis.

Our findings have already demonstrated some intriguing differences between commensal and pathogenic milk bacteria with potential role in host-microbe interaction. Further results will be discussed.

References

1. Savijoki K, H Ingmer, D Frees, FK Vogensen, A Palva and P Varmanen. 2003. Heat and DNA damage induction of the LexA-like regulator, HdiR from *Lactococcus lactis* is mediated by RecA and ClpP. Molecular Microbiology 50:609-621.
2. Varhimo E, K Savijoki, J Jalava, O Kuipers, and P Varmanen. 2007. Identification of a novel streptococcal gene cassette mediating SOS-mutagenesis in *Streptococcus uberis*. Journal of Bacteriology 89:5210-5222.
3. Prudhomme M, L Attaiech, G Sanchez, B Martin B, JP Claverys. 2006. Antibiotic stress induces genetic transformability in the human pathogen *Streptococcus pneumoniae*. Science. 313:89-92.
4. Siljamäki P, A Iivanainen, PK Laine, M Kankainen, H Simojoki, T Salomäki, S Pyörälä, L Paulin, P Auvinen, K Koskinen, L Holm, T Karonen, S Taponen, TA Nyman, A Sukura, N Kalkkinen, K Savijoki, P Varmanen. Multigenome analysis and proteome profiling of *Staphylococcus epidermidis* strains to screen for host-specificity determining factors, *in preparation*.
5. Siljamäki P, T Karonen, K Koskenniemi, TA Nyman, K, P Varmanen, K. Savijoki. Two-dimensional difference gel electrophoresis to explore strain-dependent proteomes of *Staphylococcus epidermidis* of human and bovine origin, *in preparation*.
6. Siljamäki P, K Savijoki, N Lietzen, P Varmanen, M Kankainen, TA Nyman. Comparative Exoproteomics of *Staphylococcus epidermidis* of human and bovine origin to identify bacterial factors involved in adaptation into bovine host, *in preparation*.
7. Suokko A, M Poutanen, K Savijoki, N Kalkkinen, P Varmanen. 2008. ClpL is essential for induction of thermotolerance and is potentially part of the HrcA regulon in *Lactobacillus gasseri*. Proteomics. 8:1029-1041.
8. Koskenniemi K, J Koponen, M Kankainen, K Savijoki, S Tynkkynen, WM de Vos, N Kalkkinen, P Varmanen. 2009. Proteome analysis of *Lactobacillus rhamnosus* GG using 2-D DIGE and mass spectrometry shows differential protein production in laboratory and industrial-type growth media. J. Proteome Res. 8:4993-5007.
9. Savijoki K, Lietzen N, Kankainen M, Alatossava T, Koskenniemi K, Varmanen P, and TA Nyman. 2011. Comparative proteome cataloging of *Lactobacillus rhamnosus* strains GG and Lc705. Journal of Proteome Research 5:3460-3473.

10. Koskenniemi K, Laakso K, Koponen J, Kankainen M, Greco D, Auvinen P, Savijoki K, Nyman TA, Surakka A, Salusjärvi T, de Vos WM, Tynkkynen S, Kalkkinen N and P Varmanen. 2011. Proteomics and transcriptomics cha racterization of bile stress response in probiotic *Lactobacillus rhamnosus* GG. Molecular & Cellular Proteomics, 10:M110.002741.

11. Laakso K, Koskenniemi K, Koponen J, Kankainen M, Surakka A, Salusjärvi T, Tynkkynen S, Savijoki K, Nyman TA, Kalkkinen N, Tynkkynen S, and P Varmanen. 2011. Proteome and transcriptome analyses of log to stationary growth phase *Lactobacillus rhamnosus* GG cells, *manuscript.* Microbial Biotechnology 4:746-766.

12. Koponen J, Laakso K, Koskenniemi K, Kankainen M, Surakka A, Savijoki K, Nyman TA, Salusjärvi T, Tynkkynen S, de Vos WM, Kalkkinen N, and P Varmanen. 2012. Effect of pH on proteome and transcriptome of *Lactobacillus rhamnosus* GG. Journal of Proteomics. 75:1357-1374.

13. Poutanen M, E Varhimo, N Kalkkinen, A Sukura, P. Varmanen and K Savijoki. 2009. Two-dimensional difference gel electrophoresis analysis of *Streptococcus uberis* in response to mutagenesis-inducing ciprofloxacin challenge. Journal of Proteome Research 8:246-255.

14. Varhimo E, Varmanen P, Fallarero A, Skogman M, Pyörälä S, Vuorela P, Iivanainen A, Sukura A, and K Savijoki. 2011. The α- and β-casein components of host milk induce biofilm formation in the bovine mastitis pathogen *Streptococcus uberis.* Veterinary Microbiology 149:381-389.

15. Varhimo E, K Savijoki, H Jefremoff, J Jalava, A Sukura and P Varmanen. 2008. Ciprofloxacin induces mutagenesis to antibiotic resistance independent of UmuC in *Streptococcus uberis.* Environmental Microbiology 10:2179-2183.

Tick-borne diseases in cattle: applications of proteomics and the development of new generation vaccines

Isabel Marcelino[1,2,3]*, André Martinho de Almeida[2,4]*, Miguel Ventosa[2,3], Ludovic Pruneau[1], Damien F. Meyer[1], Dominique Martinez[5], Thierry Lefrançois[1], Nathalie Vachiéry[1] and Ana Varela Coelho[2]

[1]CIRAD, UMR CMAEE, F-97170 Petit-Bourg, Guadeloupe, FWI; isabel_m31@hotmail.com
[2]ITQB-UNL, Av. da República, Estação Agronómica Nacional, 2780-157 Oeiras, Portugal
[3]IBET, Apartado 12, 2780-901 Oeiras, Portugal
[4]IICT/CVZ & CIISA, FMV, Av. Univ. Técnica, 1300-477 Lisboa, Portugal
[5]CIRAD, UMR CMAEE, F-34398 Montpellier, France
* These authors contributed equally to this work.

Introduction

Tick-borne diseases (TBDs) affect 80% of the world's cattle population, severely hampering livestock production. Some of these diseases are caused by obligate intracellular tick-borne pathogens (TBPs) such as *Theileria* spp., *Babesia* spp., *Anaplasma marginale* and *Ehrlichia ruminantium* (1). Immunization strategies against TBDs are currently available (most of them being blood-derived or attenuated), but with variable efficacy (2); thus identification of new antigens is required to develop improved vaccines. Nowadays, new breakthroughs in vaccine research are increasingly reliant on novel 'Omics' approaches such as genomics, proteomics, transcriptomics, and metabolomics, to deepen our understanding of the key biological processes that lead to protective immunity (Figure 1). Despite the recent availability of TBPs genomes, the increased knowledge on their biology has proven to be difficult since they have complex life cycles (with different developmental forms) being able to infect and multiply within different types of cells. As limited genetic tools are currently available to complement existing genomic information, post-genomics strategies such as proteomics are nowadays being used more frequently.

Results and discussion

Available proteomic studies on the above mentioned TBDs are either related to the pathogen, infected host cells and/or in vivo expression analysis of the tick vector.

For instance, in *Theileria* spp., 2D electrophoresis and MALDI-TOF was used to analyze the protein expression pattern of different *T. parva* strains (3) and the schizont-protein spot patterns of the same *Theileria* stabilate cultivated in two different infected cell lines. While no significant difference in protein expression pattern was observed between strains, the author observed a differential protein expression pattern depending on host cells. To detect

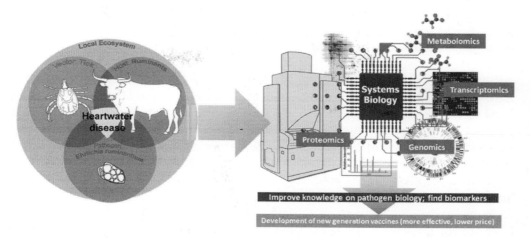

Figure 1. Omics and vaccine discovery. The development of new vaccines against TBDs such as Heartwater, imply implies the profound knowledge of the intimate relations between host-vector-pathogen. Vaccine candidates can be identified by analysis of the pathogen's genome, transcriptome, proteome, or metabolome. Systems biology will integrate the overall data to build prediction methods to identify protective epitopes. (artwork by Simão Mateus)

immunodominant schizont surface proteins (surfome), *Theileria* spp. surface antigens were characterized using 2D and Western blot (4). As *Theileria* spp. transform host cell (leucocytes) inducing uncontrolled proliferation, differential protein expression patterns in infected bovine lymphoblastoid cells was also performed (5); the results showed that ten proteins were found in infected cells but not in uninfected cells, and seven of these were detected in preparations of purified schizonts.

Most studies available on *Babesia* spp. are from tick cell models. The identification of these proteins may provide new insights into the molecular interactions between *Babesia* spp. and the tick vector. Using 2D followed by MS or/and capillary-HPLC-electrospray tandem mass spectrometry (HPLC-ESI-MS/MS), differences in expression of proteins from ovaries (6) and midgut tissue (7) of infected versus non-infected *Rhipicephalus. (Boophilus) microplus* ticks were assessed. The results showed that infection with *Babesia* induced up-regulation of five metabolic enzymes related to electron and proton transport, protein processing and retinoic acid metabolism.

Proteomic studies on *A.marginale* have been mainly targeted to identify outer membrane proteins (OMPs), as these proteins are known to induce protective immune response in cattle and to understand the transition from the host to the tick vector. Several studies were performed to analyze differentially regulated proteins between *A.marginale*-infected and uninfected host and tick cells (8,9). These studies allowed identifying several immunogenic proteins from the

OMP family and from the Type Four Secretion System; vaccination experiments were used to validate their use as potential vaccine candidate. Proteomics studies also revealed that *A.marginale* surfome in vector cells is less complex that the one in host cells (9).

Few proteomics studies are currently available for *E. ruminantium*, being mainly related to the differential expression of the immunodominant *E. ruminantium* MAP1 family proteins in infected bovine endothelial (10,11) and tick cell cultures (11). These studies revealed that this MAP1 protein is differentially expressed not only in both tick and host cells but also along the developmental cycle of the bacterium inside the host cells. In 2011, our group presented the first partial proteome map of *E.ruminantium* (10); 25% of the identified were found to be isoforms, revealing that post-translational modifications can be of significant importance for *E.ruminantium*.

Interestingly, for some TBPs, more transcriptomics studies are available compared to proteomics data. This is the case for *E. ruminantium*; indeed several manuscripts described the global gene expression profiling of *Ehrlichia* at different stages of development or are specifically related to *map1* gene expression either in host or vector cells (12-14). Whole transcriptome analysis were also performed for *Theileria* spp. in host cells, namely from two breeds of cattle (resistant and susceptible to disease) but also during the complex developmental cycle (15,16). In *Babesia* spp., transcriptomics studies were performed to study antigenic variation but also to study peptidases expression in attenuated versus virulent strain (17,18); an expression oligonucleotide microarray was also developed (19). For *A.marginale*, differential gene expression was studied in different tick and host cells gene expression in different tick and host cells (20,21).

Concluding remarks

From the results presented above, it is clear that the use of 'Omics' is still in its infancy regarding TBDs. Still, we strongly believe that these high-throughput cutting edge technologies will now significantly contribute to overcome knowledge gaps on the role of key parasite molecules involved in cell invasion, adhesion, tick transmission and, surely revolutionize the length of time and capacity for discovering potential candidate vaccines (such as proteins involved in protective immune response, tick feeding, or parasite development). By joining all the Omics data using an integrative systems biology approach, we also expect to possibly be able to develop a new vaccine against several TBDs to simultaneously avoid the transmission of tick-borne diseases and control tick infestations.

Acknowledgements

Authors acknowledge funding from project ER-TRANSPROT (PTDC/CVT/114118/2009), Post-doc grant SFRH/ BPD/ 45978/ 2008 (I. Marcelino) and the Ciência 2007 program (AM Almeida), all by Fundação para a Ciência e a Tecnologia (Lisboa, Portugal).

References

1. B. Minjauw, A. McLeod, Tick-borne diseases and poverty. The impact of ticks and tick- borne diseases on the livelihood of small-scale and marginal livestock owners in India and eastern and southern Africa. (Centre for Tropical Veterinary, Medicine, University of Edinburgh, UK., 2003).

2. V. Shkap, A. J. de Vos, E. Zweygarth, F. Jongejan, Trends Parasitol 23, 420-6 (Sep, 2007).

3. C. Sugimoto *et al.*, Parasitol Res 76, 1-7 (1989).

4. C. Sugimoto *et al.*, Parasitol Res 78, 82-5 (1992).

5. C. Sugimoto *et al.*, Mol Biochem Parasitol 37, 159-69 (Dec, 1989).

6. A. Rachinsky, F. D. Guerrero, G. A. Scoles, Insect Biochem Mol Biol 37, 1291-308 (Dec, 2007).

7. A. Rachinsky, F. D. Guerrero, G. A. Scoles, Vet Parasitol 152, 294-313 (Apr 15, 2008).

8. J. E. Lopez *et al.*, Infect Immun 73, 8109-18 (Dec, 2005).

9. S. M. Noh *et al.*, Infect Immun 76, 2219-26 (May, 2008).

10. I. Marcelino *et al.*, Vet Microbiol (Dec 1, 2011).

11. M. Postigo *et al.*, Vet Microbiol 128, 136-47 (Apr 1, 2008).

12. L. Pruneau *et al.*, FEMS Immunol Med Microbiol (Nov 18, 2011).

13. C. P. Bekker *et al.*, Gene 285, 193-201 (Feb 20, 2002).

14. M. Postigo *et al.*, Vet Microbiol 122, 298-305 (Jun 21, 2007).

15. C. A. Oura, S. McKellar, D. G. Swan, E. Okan, B. R. Shiels, Cell Microbiol 8, 276-88 (Feb, 2006).

16. K. Jensen *et al.*, Int J Parasitol 38, 313-25 (Mar, 2008).

17. B. Al-Khedery, D. R. Allred, Mol Microbiol 59, 402-14 (Jan, 2006).

18. M. Mesplet *et al.*, Mol Biochem Parasitol 179, 111-3 (Oct, 2011).

19. A. O. Lau, D. L. Tibbals, T. F. McElwain, Exp Parasitol 117, 93-8 (Sep, 2007).

20. J. T. Agnes *et al.*, Infect Immun 78, 2446-53 (Jun, 2010).

21. R. F. Mercado-Curiel, G. H. Palmer, F. D. Guerrero, K. A. Brayton, Int J Parasitol 41, 851-60 (Jul, 2011).

The applications of proteomics to animal models of leptospirosis: *in vivo veritas*

Jarlath E. Nally
School of Veterinary Medicine, University College Dublin, Dublin 4, Ireland;
jarlath.nally@ucd.ie

Leptospirosis is a bacterial zoonotic disease with a worldwide distribution. This neglected disease is a significant public health problem, and is increasingly being recognized in developing countries and tropical regions. Major outbreaks of leptospirosis are associated with flooding after severe weather and the rapid urbanization in developing countries where slum settlements have produced the conditions for epidemic rat-borne transmission of the disease. Understanding how *Leptospira* infects humans and animals is of fundamental importance for the development of effective control strategies. Animal models significantly advance studies to elucidate pathogenic mechanisms of leptospirosis since they study the pathogenesis of infection as it actually occurs; *in vivo* veritas. Leptospires are inherently difficult bacteria to work with: they have fastidious growth requirements and cultures become avirulent as they are maintained in the laboratory. Cultures are maintained at 28-30 °C which differs significantly from the body temperature of 37 °C encountered during infection of mammals. Leptospires react to this changing environmental signal and others, including changes in pH, osmolarity, iron concentrations, and presence of serum, by regulating gene and protein expression to facilitate disease. Thus, in order to identify how leptospires adapt to the host during acute and chronic infection, animal models of chronic and acute disease have been developed. Clinically relevant pathogenic isolates of *Leptospira interrogans* serovar Copenhageni were used to

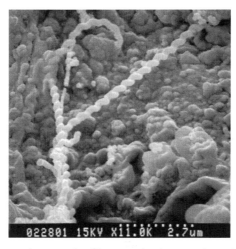

Figure 1. Scanning electron micrograph of Leptospira *interacting with equine conjunctiva.*

develop acute and chronic disease models of infection in guinea pigs and rats respectively. Results confirm that leptospires regulate protein expression during infection and that they differ significantly from their *in vitro* cultivated counterparts (Monahan *et al.*, 2008; Nally *et al.*, 2005, 2007).

Acute infection

Motile leptospires were purified from guinea pig liver by centrifugation on Percoll density gradients and compared to Percoll-purified *in vitro*-cultivated *Leptospira* (IVCL). The lipopolysaccharide O antigen (Oag) content of guinea pig liver-derived leptospires was markedly reduced compared to that of IVCL, as demonstrated both by immunoblotting with a monoclonal antibody that was serovar specific for Oag and by periodate-silver staining. In addition, the proteome of *Leptospira* expressed during actual disease processes was characterized relative to that of IVCL after enrichment for hydrophobic membrane proteins with Triton X-114. Protein samples were separated by two-dimensional gel electrophoresis, and antigens expressed during infection were identified by immunoblotting with monospecific antiserum and convalescent rat serum in addition to mass spectrometry. Results suggest a significant increase in the expression of the outer membrane protein Loa22 during acute infection of guinea pigs relative to other outer membrane proteins, whose expression is generally diminished relative to expression in IVCL. Significant amounts of LipL32 are also expressed by *Leptospira* during acute infection of guinea pigs.

Chronic infection

Rats and dogs are natural carriers of *Leptospira*, and act as reservoirs for transmission of disease to humans (Bonilla-Santiago and Nally, 2011; Rojas *et al.*, 2010). Experimentally infected rats remained clinically asymptomatic but shed leptospires in urine for several months at concentrations of up to 10^7 leptospires/ml of urine. Proteomic analysis of rat urine-isolated leptospires compared to *in vitro* cultivated leptospires confirmed differential protein and antigen expression, as demonstrated by two-dimensional gel electrophoresis and immunoblotting. Furthermore, while serum from chronically infected rats reacted with many antigens of IVCL, few antigens of rat urine-isolated *Leptospira* were reactive. Loa22, a virulence factor of *Leptospira*, as well as the GroEL, were increased in leptospires excreted in urine compared to *in vitro* cultivated leptospires. Differentially expressed host urinary proteins include Neprilysin, napsin A aspartic peptidase, and immunoglobulin G and A. Results confirm differential protein expression by both host and pathogen during chronic disease and include markers of kidney function and immunoglobulin which are potential biomarkers of infection (Nally *et al.*, 2011).

Characterization of those host-pathogen factors that facilitate acute and chronic disease processes will progress the understanding of the biology of *Leptospira* to improve diagnostics and the prevention of this neglected emerging disease.

References

Bonilla-Santiago, R. & Nally, J. E. 2011. Rat model of chronic leptospirosis. Curr Protoc Microbiol, Chapter 12, Unit12E 3.

Monahan, A. M., Callanan, J. J. & Nally, J. E. 2008. Proteomic analysis of Leptospira interrogans shed in urine of chronically infected hosts. Infect Immun, 76, 4952-8.

Nally, J. E., Chantranuwat, C., Wu, X. Y., Fishbein, M. C., Pereira, M. M., Da Silva, J. J., Blanco, D. R. & Lovett, M. A. 2004. Alveolar septal deposition of immunoglobulin and complement parallels pulmonary hemorrhage in a guinea pig model of severe pulmonary leptospirosis. Am J Pathol, 164, 1115-27.

Nally, J. E., Chow, E., Fishbein, M. C., Blanco, D. R. & Lovett, M. A. 2005. Changes in lipopolysaccharide O antigen distinguish acute versus chronic Leptospira interrogans infections. Infect Immun, 73, 3251-60.

Nally, J. E., Monahan, A. M., Miller, I. S., Bonilla-Santiago, R., Souda, P. & Whitelegge, J. P. 2011. Comparative proteomic analysis of differentially expressed proteins in the urine of reservoir hosts of leptospirosis. PLoS One, 6, e26046.

Nally, J. E., Whitelegge, J. P., Bassilian, S., Blanco, D. R. & Lovett, M. A. 2007. Characterization of the outer membrane proteome of Leptospira interrogans expressed during acute lethal infection. Infect Immun, 75, 766-73.

Rojas, P., Monahan, A. M., Schuller, S., Miller, I. S., Markey, B. K. & Nally, J. E. 2010. Detection and quantification of leptospires in urine of dogs: a maintenance host for the zoonotic disease leptospirosis. Eur J Clin Microbiol Infect Dis, 29, 1305-9.

Part II
Proteomics and animal health

Changes in the proteome of the H295R steroidogenesis model associated with exposure to the mycotoxin zearalenone and its metabolites, α- and β-zearalenol

Øyvind L. Busk[1], Doreen Ndossi[2,3], Caroline Frizzell[4], Steven Verhaegen[2], Silvio Uhlig[5], Gunnar Eriksen[5], Lisa Connolly[4], Erik Ropstad[2] and Morten Sørlie[1]
[1]*Department of Chemistry, Biotechnology and Food Science, Norwegian University of Life Sciences, P.O. Box 5003, 1432 Ås, Norway*
[2]*Norwegian School of Veterinary Science, Oslo, Norway; steven.verhaegen@nvh.no*
[3]*Sokoine University of Agriculture, Tanzania*
[4]*Institute of Agri-Food and Land Use, School of Biological Sciences, Queen's University Belfast, Ireland*
[5]*Section for Chemistry and Toxicology, Norwegian Veterinary Institute, Oslo, Norway*

Mycotoxins are secondary fungal metabolites that can cause pathological, physiological, and/or biochemical alterations in other species, including higher vertebrates. Zearalenone (ZEN) is produced by the genus *Fusarium* and after ingestion via contaminated cereals, may lead to animal fertility problems and reproductive pathologies. The compound has endocrine disrupting effects with a strong estrogenic action. Current data indicates that pig (hyperestrogenism) is amongst the most susceptible to ZEN, whilst in humans, this mycotoxin is suspected to be a triggering factor for central precocious puberty development in girls. Biotransformation of ZEN forms the metabolites α-zearalenol (α-ZOL) and β-zearalenol (β-ZOL), which are also produced by *Fusarium* in cultures, but in lower concentrations than ZEN.

The human adrenal carcinoma cell line, H295R, is capable of full hormone production and thus forms an exquisite *in vitro* model to study steroidogenesis. As such, the H295R steroidogenesis assay has been established as validated model for endocrine disruption (Hecker *et al.*, 2011). The effects of ZEN and its metabolites, α-ZOL and β-ZOL, have previously been investigated at the level of nuclear receptor transcriptional activity, using a panel of mammalian reporter gene assays (RGAs), incorporating natural steroid receptors, and on steroidogenesis, using the human adrenal carcinoma cell line, H295R (Frizzell *et al.*, 2011).

We recently analysed changes induced in the proteome of this cell line, following exposure to 10 μM of each of the 3 compounds for 48 hrs, using a metabolic labelling approach: *S*table-*I*sotope *L*abelling by *A*mino acids in *C*ell culture (SILAC). Cell viability was not affected at this concentration. Following exposure, proteins were extracted from four different subcellular fractions. Isolated protein fractions were run on a SDS-Page gel, each lane was divided into 12 gel pieces, and the pieces digested with trypsin. Peptides isolated from these pieces were identified using a mass spectrometer/mass spectrometer system.

For ZEN a total of 21 differentially regulated proteins were identified in the four fractions (Table 1). System biology analysis, using Ingenuity Pathway Analysis, indicated several proteins connected with estrogenicity, like e.g. several regulated cytochrome-C oxidase proteins, along with several other cytochrome proteins and serotransferrin (Table 1). Another network centers around the transcription factor Nuclear Factor kappa B (NFkB), which responds to a number of different stressors, and that controls a number of important cellular functions. Canonical pathways involved include the mitochondrial dysfunction pathway and the oxidative phosphorylation pathway (Busk *et al.*, 2011).

For α-ZOL and β-ZOL a total of 14 and 5 individual proteins in the cytosolic fraction were found to be significantly regulated, respectively. For α-ZOL, the observed regulated proteins are connected in a network related to cellular movement, cell-to-cell signalling and interaction,

Table 1. List of identified proteins regulated by zearalenone in a H295R model cell line.

Protein names	Gene names	Ratio H\L	Subc. fraction
Activating molecule in BECN1-regulated autophyagy protein 1	AMBRA1	0.084å	4
Ankyrin repeat domain-containing protein 27	ANKRD27	4.693	1
Cytochrome b-c1 complex subunit 1, mitochondrial	UQCRC1	0.660	3 2
Cytochrome b-c1 complex subunit 2, mitochondrial	UQCRC2	0.643	3 2
Cytochrome c oxidase subunit 2	MT-CO2	0.645	3 2
Cytochrome c oxidase subunit 4 isoform 1, mitochondrial	COX4I1	0.635	3 2
Cytochrome c oxidase subunit 5B, mitochondrial	COX5B	0.611	2
DNA-binding protein RFX7	RFX7	0.215	1
Heat shock protein 90kDa beta member 1	GRP94c	0.089	1
Hepatoma-derived growth factor	HDGF	0.571	1 2 3
Lymphoid-restricted membrane protein	LRMP	0.041	1
Mitochondrial import receptor subunit TOM40 homolog	TOMM40	0.640	2 3
Patched domain-containing protein C6orf138	C6orf138	16.896	1
Protein Shroom3	SHROOM3	0.103	4
Annexin A2	ANXA2	0.206	4
Ribosomal protein L29	RPL29	0.444	2
Scavenger receptor class B member 1	SCARB1	1.364	1 2 3 4
Serotransferrin	TF	0.126	1 2 3 4
Synembryn-B	RIC8B	0.093	1
Transcription factor BTF3 homolog 4	BTF3L4	0.656	1
Trypsin-3	PRSS3	0.150	4

Proteins identified with a corrected *P*-value <0.01

cellular growth and proliferation (Figure 1), while for β-ZOL it was observed to be cellular growth and proliferation, hematological system development and function, and tissue morphology (Busk *et al.*, 2012).

Interestingly, there are no common protein regulations by the three individual mycotoxins. This lack of similarity in the affected genes from the three different inductions could be a consequence of the structural difference of the metabolites. However, it cannot be ruled out

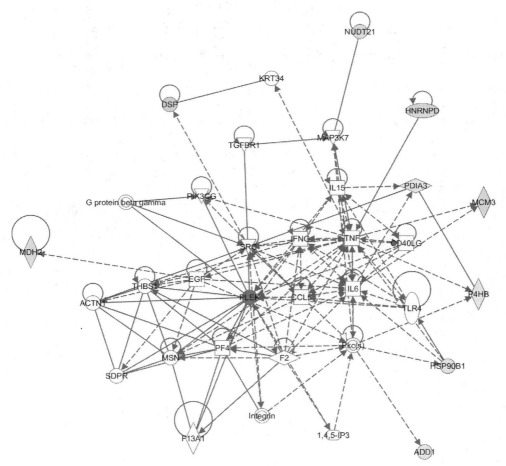

Figure 1. Molecular relationships between molecules (protein interaction network) found to be regulated by α-ZOL. The network was generated using Ingenuity Pathway (IPA) (red=up-regulated, green=down-regulated; strength of colour = the amount of regulation). White nodes are proteins not included in the input set of proteins but are used to connect the regulated proteins. Adapted from Busk et al., 2012 (supplementary material).

that the complexity of the samples is high preventing the detection of significantly regulated proteins that are similar for the individual mycotoxins.

Taken together the data indicate that non-cytotoxic concentrations of ZEN, and its metabolites α-ZOL and β-ZOL, can modify the proteome of these cells, and as such can potentially modulate the biological functions of adrenal cells. These subtle changes can be detected by our proteomics approach, and in the future we intend to apply this method to an in-house model for steroidogenesis based on primary Leydig cells isolated from pig testicles.

References

Busk ØL, Frizzell C, Verhaegen S, Uhlig S, Connolly L, Ropstad E, Sørlie M. Cytosol protein regulation in H295R steroidogenesis model induced by the zearalenone metabolites, α- and β-zearalenol. Toxicon. 2012 Jan;59(1):17-24.

Busk ØL, Ndossi D, Verhaegen S, Connolly L, Eriksen G, Ropstad E, Sørlie M. Relative quantification of the proteomic changes associated with the mycotoxin zearalenone in the H295R steroidogenesis model. Toxicon. 2011 Nov;58(6-7):533-42.

Frizzell C, Ndossi D, Verhaegen S, Dahl E, Eriksen G, Sørlie M, Ropstad E, Muller M, Elliott CT, Connolly L. Endocrine disrupting effects of zearalenone, alpha- and beta-zearalenol at the level of nuclear receptor binding and steroidogenesis. Toxicol Lett. 2011 Oct 10;206(2):210-7.

Hecker M, Hollert H, Cooper R, Vinggaard AM, Akahori Y, Murphy M, Nellemann C, Higley E, Newsted J, Laskey J, Buckalew A, Grund S, Maletz S, Giesy J, Timm G. The OECD validation program of the H295R steroidogenesis assay: Phase 3. Final inter-laboratory validation study. Environ Sci Pollut Res Int. 2011 Mar;18(3):503-15.

Proteomic analysis of host responses to *Salmonella enterica* serovar Typhimurium infection in gut of naturally infected Iberian pigs

Cristina Arce, Angela Moreno and Juan Jose Garrido
Grupo de Genómica y Mejora Animal, Departamento de Genética, Facultad de Veterinaria, Universidad de Córdoba-CSIC, Campus de Rabanales, 14071 Córdoba, Spain; ge1gapaj@uco.es

Salmonella Typhimurium is the serotype most frequently isolated from ill pigs in Europe. *Salmonella* outbreaks and subclinical infections are often not only a cause of economic and animal welfare costs, but also a source of contamination of pork products entering the food chain. Currently, the implementation of animal breeding projects, as well as the design of effective vaccines are considered the base for effective and sustainable disease control. Therefore, a better understanding of host responses against *Salmonella* infection at the mucosal barrier is crucial to achieve this objective. Although some successful physiological, biochemical or genetic approaches have been developed at intestinal level contributing with new insights, no proteomic investigations have been performed until now. *S.* Typhimurium infection and pathogenicity are complex, highly integrated processes that cannot be attributed to any single protein activity. However, advances in proteomic technologies now make it possible to characterize host-pathogen interactions from a global proteomic point of view. In this study, we used a proteomic approach to study proteins of intestinal mucosa differentially expressed as response to *Salmonella* infection

Material and methods

Three healthy weaned Iberian pigs and three animals naturally infected with *Salmonella* Typhimurium were selected to be used as control or infected animals. Frozen samples from small gut were rapidly thawed. Sections of 4-5 cm length were cut and luminal surface was thoroughly cleaned using sterile gauze and PBS to eliminate mucus and blotted dry onto dried gauze. Intestine mucosa sections were scraped with a razor and resuspended in a lysis buffer (7 M urea, 2 M thiourea, 4% CHAPS, 30 mM Tris, buffered to pH 8). After clean up precipitation, protein samples were solubilized in 2-D DIGE sample buffer (7 M urea, 2 M thiourea, 4% CHAPS, buffered to pH 8.5 with NaOH).

Samples were cup-loaded onto IPG strip (24 cm, pH 3-11 NL) and subjected to IEF. For the second dimension, strips were loaded on SDS-PAGE (12,5%). Gels were scanned using a Typhoon Trio Imager and analyzed with Decyder 6.5 software. Spots were excised automatically in a ProPic station and subjected to MS analysis by MALDI-TOF/TOF or LC-ESI-MS/MS.

Results and discussion

In order to identify potential biomarkers of salmonelosis in pigs, we have utilized a proteomic strategy in attempts to identify gut proteins involved in the response of Iberian pigs naturally infected with *Salmonella* Typhimurium. Previous studies have evidenced differences between findings in experimentally and naturally infected wildlife species (Naranjo *et al.*, 2006), thus stressing the importance of studies in naturally infected animals. 2D-DIGE was used to identify differential mucosal protein expression in the small intestine of Iberian pigs as a consequence of the *Salmonella* Typhimurium infection. Differences in spots intensity among gels from control and infected samples were identified in 71 matched spots. Forty-nine spots were identified, 28 by MALDI-TOF/TOF and 21 by LC-ESI-MS/MS, as differentially regulated (*P*<0.05), corresponding to 42 different proteins, with 12 proteins showing up-regulation and 30 down-regulation (Figure 1). Analysis of results according to Gene Ontology categories of subcellular localization revealed the presence of proteins from different origins, including nucleus, cytoskeleton and cytoplasm. These proteins were involved in different biological processes and pathways, including acute phase response (RBP2, APOA1); immunoregulatory proteins

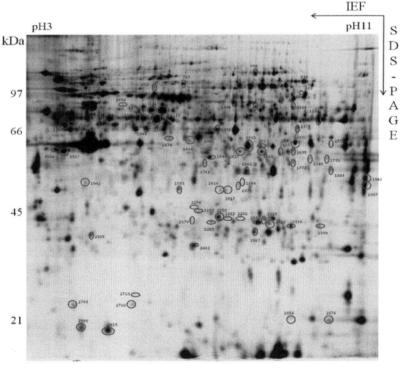

*Figure 1. Master gel obtained after analysis by the DeCyder software. 49 spots exhibited differences (*P<0.05*).*

(APOA1, ANXA4) or cytoskeleton related proteins (KRT8, KRT19, KRT10, EZR, ACTR3, ARCP5). The association of the deregulated proteins with canonical pathways highlighted three major pathways: glycolysis/glucogenesis, regulation of actin-based motility by Rho and actin cytoskeleton signaling. These results are in agreement with previously reported data indicate that *Salmonella* Typhimurium requires glycolysis for infection of mice and macrophages and that transport of glucose is required for replication within macrophages (Bowden *et al.*, 2009). Also has been observed that the actin cytoskeleton rearrangements, which lead to *Salmonella* internalization, are achieved by the activation of host GTPases of the Rho family.

The induction of host-cytoskeleton modifications by *Salmonella* effectors has been a subject of extensive study (Bakowski *et al.*, 2008). In our study, a variable regulation occurred for cytoskeleton structural proteins and for cytoskeleton assembly, dynamics and signalling proteins such as ezrin and ARCP5. The complexity of the tissue under study made difficult a direct association among the expression changes of cytoskeletal proteins and a concrete cellular function. However, some of the observed regulations could be a direct consequence of the *Salmonella* Typhimurium presence in the tissue. Thus, related to the interaction with non-immune cells would be the down-regulation of ezrin (Friederich *et al.*, 1999) required for *Salmonella* Typhimurium entrance in epithelial cells (McGhie *et al.*, 2009). Moreover, modifications of cytoskeleton proteins could also be consequence of *Salmonella*-containing vacuole (SCV) dynamics, the intracellular structure for survival and replication in host cells, since this is dictated by a fine balance of actin and microtubule motors (Bakowski *et al.*, 2008).

We have also found down-expressed the expression of caspase-1, a protein with a relevant role in the inflammation process. It has been described that caspase-1 deficiency affects the course of the *S.* Typhimurium infection at least in three ways: during the initiation of inflammation, the amplification of pro-inflammatory signalling within the infected mucosa tissue and by restricting pathogen spread/growth within host tissues like the mesenteric lymph nodes, spleen and liver (Kaiser *et al.*, 2012).

When functional networks were analyzed, we found that the set of differentially expressed proteins were integrated into six networks displaying a significant score. The most significant functions associated with these networks were cell assembly and organization (Figure 2), confirming the results mentioned above.

In conclusion, proteomic study presented in this work show that, at intestinal mucosa level, the physiological functions most significantly perturbed as consequence of *Salmonella* Typhimurium infection were immunological, infectious and gastrointestinal disease, cellular assembly and organization, tissue morphology, cell death and immune response. Our results contribute to the understanding of host-pathogen interactions during *Salmonella* infections and expand the existing information on the response of the porcine gut to the bacterial infections under natural conditions.

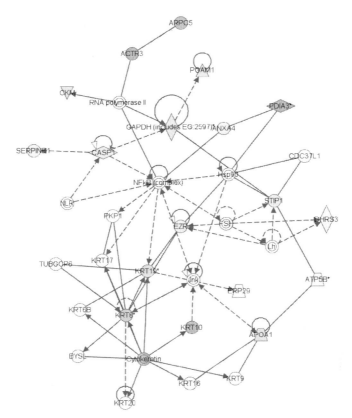

Figure 2. Ingenuity Pathway Analysis using the Ingenuity knowledge database for 18 of the 49 identified proteins. These proteins are involved in the top-one network: 'Cellular assembly and organization'. A graphical representation of the interaction network is shown, indicating proteins whose expression levels appeared as being differentially regulated in our analysis. Lines indicate protein-protein interactions.

References

Bakowski, M.A., Braun, V., Brumell, J.H., 2008. *Salmonella*-Containing Vacuoles: Directing Traffic and Nesting to Grow. Traffic 9, 2022-2031. McGhie, E.J., Brown, L.C., Hume, P.J., Humphreys, D., Koronakis, V., 2009. Curr Opin Microbiol 12, 117-124.

Bowden SD, Rowley G, Hinton JC, Thompson A., 2009. Infect Immun., 77, 3117-26.

Friederich, E., Vancompernolle, K., Louvard, D., Vandekerckhove, J., 1999. J. Biol. Chem. 274, 26751-26760.

Kaise, P., Diard, M., Stecher, B., Hartd, W., 2012. Immunol. Reviews. 245, 56- 83.

Naranjo, V., Ayoubi, P., Vicente, J., Ruiz-Fons, F., Gortazar, C., Kocan, K.M., de la Fuente, J., 2006. Vet.Microb. 116, 224-231.

Biochemical and proteomic investigation of bovine nasal secretion

M. Faizal Ghazali[1], Nicholas Jonsson[1], Richard J.S. Burchmore[2] and P. David Eckersall[1]
[1]School of Veterinary Medicine, University of Glasgow, Bearsden Rd, G61 1QH, Glasgow, United Kingdom; m.ghazali.1@research.gla.ac.uk
[2]Glasgow Polyomics Facility, University of Glasgow, G12 8QQ, Glasgow, United Kingdom

Introduction

Disease of the respiratory tract in cattle, apart from having animal welfare and potential human health implications, causes substantial economic loss in the cattle industry every year (Gunn and Stott, 1998). Despite the widespread use of vaccines and antibiotics to combat these diseases, it still remains the most costly endemic disease in calves seen after the neonatal period. The clinical biochemistry of bovine nasal secretion has not been well characterised but is likely to provide useful indicators of immune function and status. The aim of the present study was to establish a viable collection system for normal nasal secretion from cattle with minimal invasive procedures and to use this to enable a proteomic and biochemical analysis of normal bovine nasal secretion, with a view to identification of biomarkers for immune function of the bovine nasal mucosa.

Material and methods

Nasal secretions were collected on to absorbent material from thirty eight clinically healthy Holstein-Friesian cows aged 2-5 years on the University of Glasgow Cochno Farm and were extracted from the absorbent by centrifugation at 1000xg at 4 °C for 10 min. The nasal secretions were examined for biochemical composition using an Olympus (Olympus, Tokyo, Japan) A640 analyzer (Veterinary Diagnostic Services, University of Glasgow), protein concentration using a Bradford assay, measurement of immunoglobulin A and G concentrations using specific ELISAs (Bethyl Labs, Texas, USA). Six of the nasal secretion samples were randomly selected and pooled for separation by 2-dimension electrophoresis (DE). The 2-DE was undertaken on a Criterion system using 11 cm, pH 3-10 IPG Strips (Bio-Rad Labs, Hemel Hempstead, UK) followed by SDS-PAGE on 4-12% polyacrylamide gels and stained with Coomassie blue. The 6 samples were subsequently run on individual 2-DE gels and protein spots were excised and subjected to tryptic in-gel digestion. Samples were then were analysed on a 4700 Proteomics Analyser MALDI-TOF/TOF mass spectrometer (Applied Biosystems, Framingham, Massachusetts, USA). Analysis of the MS and MS/MS spectral data was performed using *SwissProt* database version *57.15*. Protein scores with CI.95% were considered indicative of a positive identification with individual ion scores at P=0.05 or better, indicative of identity or extensive homology.

Results

Protein concentrations in bovine nasal secretion ranged from 8.21 to 33.7 g/l. Alkaline phosphatase (ALP) and gamma-glutamyltransferase (GGT) concentrations were high compared with the reference range for serum (Table 1). Other biochemical analytes were within or near the reference range for serum. The concentrations of immunoglobulin A and immunoglobulin G in nasal secretion were between 0.45 to 1.82 g/l and 0.17 to 1.88 g/l respectively.

2-DE of pooled nasal secretion using 11 cm IPG strip (pH 3-10) produced a gel with well separated spots (Figure 1a) and justified examination of individual samples. Individual nasal secretions showed a similar spot pattern with little variation observable. Following trypsin digest and peptide mass fingerprinting of spots (Figure 1b) 7 major proteins were identified in bovine nasal secretion (Table 2). Serum proteins: albumin, fibrinogen, apolipoprotein A1s and serotransferrin were identified in the nasal secretion. In addition, lactotransferrin, an anti-bacterial protein known to occur in secretions such as milk and saliva, odorant binding protein, known to be involved in scent recognition and glutathione-S-transferase, an enzyme capable of detoxifying noxious compounds were putatively identified.

Table 1. Biochemistry and ELISA results.

Test analyte	Units	Mean	Standard deviation (s)	Median	Range	Reference (bovine serum)
ALP	U/l	1,192.76	500.89	1,236	144-2,392	20-280[a]
GGT	U/l	71.32	29.06	68.5	28-149	6.1-27[a]
Total Protein	g/l	16.02	6	15.63	8.21-33.7	52-84[a]
Ig A	g/l	1.20	0.33	1.19	0.45-1.82	0.06-1.0[b]
Ig G	g/l	0.60	0.46	0.54	0.17-1.88	6-15.1[b]

[a] Laboratory reference range.
[b] Duncan *et al.* 1972.

Conclusions

Samples of bovine nasal secretion have been collected from healthy animals in sufficient volume and quality for multiple analyses using a simple non-invasive collection method. Biochemical, immunological and proteomic results were collected and analysed to establish a baseline parameter of normal bovine nasal secretion. This will enable the development and validation of biomarkers in this fluid to study the pathophysiological responses of the host against respiratory diseases.

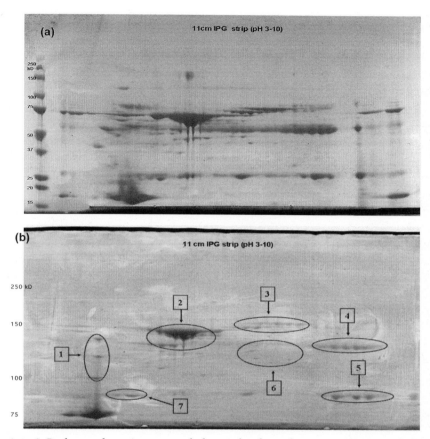

Figure 1. a: 2-D electrophoresis on a pooled sample of nasal secretion from six different cattle to provide an overview of the protein spots distribution on 2-D SDS-PAGE gel. b: Protein spots of bovine nasal secretion excised for proteomic analysis and identified by MS.

Table 2. Gel-based MS identification of nasal secretion proteins.

Protein spot no.	Protein name	Accession	Theoretical mass (Da)	Highest MS score	Highest Matches
1	Odorant-binding protein	OBP_BOVIN	18,492	498	32
2	Albumin	ALBU_BOVIN	71,244	2,568	163
3	Serotransferrin	TRFE_BOVIN	79,870	1,394	102
4	Lactotransferrin	TRFL_BOVIN	80,002	129	12
5	Glutathione S-transferase A1	GSTP1_BOVIN	23,826	525	26
6	Fibrinogen beta chain	FIBB_BOVIN	53,933	474	40
7	Apolipoprotein A1	APOA1_BOVIN	30,258	1,296	77

Acknowledgements

The Malaysian Ministry of Higher Education is thanked for funding for MFG.

References

Duncan,J.R., Wilkie,B.N., Hiestand,F. & Winter,A.J. 1972. The Journal Of Immunology. Vol. 108, No. 4 pp 965-976.

Gunn,G.J. & Stott,A.W. 1998. Proceedings XX World Buiatrics Congress, Sydney. Vol. 1, pp 357-360.

The behaviour of ceruloplasmin as an acute phase protein in obese and infected rabbits

T.M. Georgieva[1], T. Vlaykova[2], E. Dishlianova[1], Vl. Petrov[1] and I. Penchev Georgiev[1]
[1]Faculty of Veterinary Medicine, Trakia University, 6000 Stara Zagora, Bulgaria; tmgeorg@uni-sz.bg
[2]Faculty of Medicine, Trakia University, 6000 Stara Zagora, Bulgaria

Abstract

Acute phase proteins (APP) are serum proteins which increase in concentration during the acute phase reaction (APR) to inflammation or infection. The response occurs in all animals, but in different species the response of individual proteins can be significantly different. Kushner *et al*. 2006, divided positive APP in three groups: those with an increase of about 50%: ceruloplasmin and C3; those with an increase of 2 to 3 times: haptoglobin, fibrinogen and α-globulins with antiprotease activity; and those proteins with a rapid increase up to 1000-fold: C-reactive protein and serum amyloidal (SAA). Ceruloplasmin (Cp) is an APP from the α_2-globulin fraction. It is a small APP, of approximately 160 kDa, with a concentration in plasma of approximately 30 mg/dl. So far in the available literature there is no information concerning the behavior of this protein in rabbits after infection. All *S. aureus* infections in individual rabbits have a similar appearance, with lesions of pododermatitis, subcutaneous abscesses and mastitis. In rabbit flocks, two types of *S. aureus* infections can be distinguished. In the first type with a low virulent strain, the infection remains limited to a small number of animals and is of minor economic importance. In the second type with a high virulent *S.aureus*, infection results in epidemic spread of disease in the rabbitry. Rabbitries where highly virulent strains circulate, suffer substantial financial loss. The objective of this work is to compare the extent of ceruloplasmin elevation in infected obese rabbits with rabbits which were only infected.

Materials and methods

The experimental procedure was approved by the Ethic Committee at the Faculty of Veterinary Medicine.

I experiment

13 male 3 months old New Zealand white rabbits with average body weight of 3.2 kg were divided into 2 groups: infected group (n=7) and control group (n=6). The rabbits from infected group were inoculated subcutaneously with high virulent field strain *S. aureus* (0.1 ml 8×10^8 c.f.u./ml). The rabbits from control group were inoculated with saline. Blood samples with heparin were taken from *v. auricularis* before and at the 6th, 24th, 48th, 72nd hour and on days 7, 14 and 21 after inoculation.

II experiment

12 male New Zealand white, starting at 3 months of age and 2 weeks after castration, 1.5 month after growing fat, 4 months old rabbits and starting of infection. These rabbits were compared with pre-treatment period of the same animals and with controls. The mean average body weight of the rabbits at age of 4 months was 4.43±0.23 kg. Six rabbits at this age were experimentally infected subcutaneously with 100 µl of bacterial suspension of a field *S. aureus* strain (density: 8x10⁸ cfu/ml) as described by Wills *et al.* and 6 rabbits at age of 4 months served as controls. On post infection day 21 all rabbits were euthanized by injection of a lethal dose (100 mg/kg.b.w.) of Thiopental (Sandoz GmbH Austria). The carcases of the animals were subjected to gross anatomy, histopathological and bacteriological examination.

Prior to inoculation (0 h) and 6, 24, 48, 72 hours as well as the 7th day post-infection, the rectal body temperature and the presence and size of the formed abscess were recorded.

Blood samples from each rabbit from II experiment were taken into sterile heparinised tubes from *v. auricularis externa* as follow: (1) at age of 3 months, which coincided with 2 weeks after castration; (2) at age of 3,5 months and (3) at age of 4 months, which exactly coincided with 0h before and at 6 h, 24 h, 48 h, 72 h and on days 7, 14, and 21 post *S. aureus* challenge.

Cp was measured using the method of Ravin, described by Kolb and Kamishnikov, 1982, based on oxidation of p-phenylendamine. The statistical analysis of the data was performed using one way analysis of variance (ANOVA). All data are expressed as mean ± standard deviation (SD) and the differences were considered significant when p value was less than 0.05.

Results

I experiment

The results for ceruloplasmin concentration in the infected and control group are shown in Figure 1. The ceruloplasmin concentration was 279±17.5 mg/l at 0 h, elevated significantly to 561±41 mg/l at 24 h ($P<0.001$) when compared with control group and with the initial level of the experimental rabbits. This increase represented 51%. The concentration of ceruloplasmin was at the highest level at 48 h (751±39 mg/l) which represented 63% elevation and showed the strength of the injury. This significant difference between the two groups continued also at 48 h, 72 h and on day 7. On day 14 and day 21 the concentration of ceruloplasmin slowly decreased and was comparable to the initial level.

II experiment

The results for ceruloplasmin concentration in experimental (infected and obese) and for obese rabbits (controls) are shown on Figure 2. The ceruloplasmin concentration did not

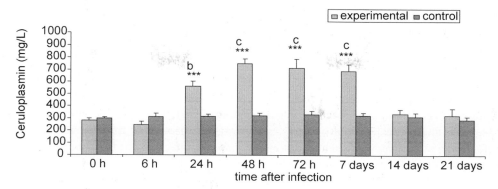

Figure 1. Concentration of ceruloplasmin in infected with St.aureus *rabbits and control rabbits in experiment I. Bars represent Mean ± SD.*

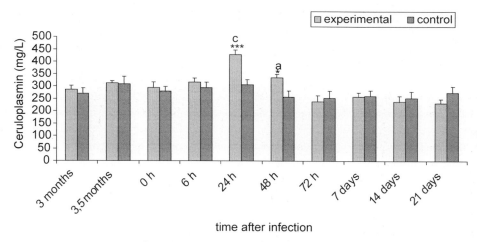

Figure 2. Concentration of ceruloplasmin in infected with St.aureus *and obese rabbits and control rabbits (only obese) in experiment II.*
Bars represent Mean ± SD.
*Significance of differences between groups: * P<0.05 *** P<0.001.*
Significance of differences in groups: [a] P<0.05; [b] P<0.05; [c] P<0.001.

differ between 3 and 4 months of age. The ceruloplasmin concentration of the rabbits from the two groups, after 1.5 month becoming fat, also did not change significantly in relation to the level of ceruloplasmin before fattening (3 months).

In the infected and obese (experimental) group, the serum ceruloplasmin rose with 32% from 293.7 mg/l at 0h of infection with *Staphylococcus aureus* to 427.6 mg/l at 24h post infection

($P<0.001$). This significant difference ($P<0.05$) continued at 48h as compared to the initial level of experimental rabbits and also to the control (only obese) group. Further, the level of ceruloplasmin slowly decreased and at 72 h, and on days 7, 14 and 21 it returned to pre-infection level in the two groups.

In conclusion, our results indicate that there was a strong and continued elevation of ceruloplasmin till day 7 only in the infected rabbits (I experiment), but not in the obese infected animals (II experiment), suggested some type of protection from the infection or a suppression of the liver acute phase response to bacterial infection due to the obesity carried out in a relatively short period of 1.5 month. The latter proposal is more likely because of the proven effect of adipose tissue and high amount of infiltrated adipose tissue macrophage (ATM) on the processes in many tissues, including liver (Lumeng and Saltiel, 2011).

References

Alsemgeest S.P.M., 1994. Blood concentration of acute phase proteins in cattle as markers for disease. PhD-thesis. Utrecht University, Utrecht, the Netherlands. ISBN 90-3-0573-0.

Devriese L.A., Godard C., Okerman L., Renault L., 1981. Characteristics of *Staphylococcus aureus* strains from rabbits. *Ann. Rech. Vét.*, 12, 3, 327-332.

Eckersall P.D., 2000. Recent advances and future prospects for the use of acute phase proteins as markers of disease in animals. *Revue Méd.Vét.*, 151, 7, 577-584.

Eckersall P.D. and J.D.Conner, 1988. Bovine and canine acute phase proteins. *Vet. Res.Communications*, 12, 169-178.

Kolb V.G., Kamishnikov V.S.: Determination of ceruloplasmin in serum by modified method of Ravina (In Russian). *In*: Practical book in clinical chemistry, Kolb V.G., Kamishnikov V.S.(eds), 2nd edition, Belaruss, 1982, pp.:290-291.

Kushner I., D.Rzeunicki, D.Samols, 2006. What does minor elevation of C-reactive protein signify?. *Amer. J.of Medicine*, 119, 166e17-166.e28.

Lumeng CN, Saltiel AR. 2011. Inflammatory links between obesity and metabolic disease. *J Clin Invest*, 121(6): 2111-7.

Okerman, L., L.Devriese, L. Maertens, F.Okerman, C.Godard, 1984. Cutaneous staphylococcosis in rabbits. *Vet.Rec.*, 114, 313-315.

Vancraevnest D., F.Haesebrouck, K.Hermans, 2007. Multiplex PCR Assay for the detection of high virulence rabbit *Staphylococcus aureus* strains. *Vet. Microbiology*, 121, 3-4, 368-372.

Wills Q.F., Kerrigan C., Soothill J.S., 2005. Experimental bacteriophage protection against *Staphylococcus aureus* abscesses in a rabbit model. *Antimicrob. Agents Chemother.*, 49 (3), 1220-1221.

Modifications of the acute phase protein haptoglobin in dairy cows

Susanne Häussler[1], Md. Mizanur Rahman[1], Andrea Henze[2], Florian Schweigert[2] and Helga Sauerwein[1]

[1]Institute of Animal Sciences, Physiology & Hygiene Unit, University of Bonn, Germany; sauerwein@uni-bonn.de

[2]Institute of Nutritional Science, University of Potsdam, Germany

Introduction

The acute phase glycoprotein haptoglobin (Hp) is mainly expressed in the liver and secreted into plasma. Hp complexes with free hemoglobin thus preventing loss of iron through the kidneys and protecting tissues from damage by hemoglobin mediated generation of lipid peroxides and hydroxyl radicals, while making the hemoglobin accessible to degrading enzymes. Recently we could confirm the relevance of Hp not only as an inflammatory marker but also as an adipokine in dairy cattle (1,2). High yielding dairy cows mobilize fat in early lactation in order to compensate the negative energy balance occurring in early lactation. In monogastric obese individuals, Hp concentrations are positively correlated with body fat content (3), however, exorbitant fat mass in high yielding dairy cows is not an issue. The expression of Hp mRNA in both subcutaneous and visceral fat depots of dairy cattle confirms the occurrence and the expression of Hp in bovine adipose tissue (2). Various site-specific covalent modifications of proteins exist, e.g. phosphorylation, glycosylation or acylation (4), regulating the configuration and the function of the proteins. A tissue-specific glycosylation has been demonstrated for many proteins and it is known that glycosylation can modify protein function or structure (5). Three different phenotypes have been classified for Hp, consisting of a combination of various α- and β-chains (6). The different phenotypes of Hp combined with altered glycosylation patterns have been reported to be associated with diseases in humans (6,7). Moreover, the type of glycosylation appears to have prognostic values in certain pathological human conditions (8). The glycosylation pattern of Hp changes, and the type of change observed can vary with disease (8). Local Hp synthesis may provide tissues with a source of functionally or structurally different Hp (5). Together with Western blot analysis, mass spectrometry can be used for analyzing the type and position of specific modifications in proteins and novel modifications can be monitored (4). We aimed to investigate the different patterns of Hp molecular weight forms present in bovine serum and tissue samples, which might be related to modified tissue specific biological functions of Hp in dairy cattle. To identify qualitatively different glycoforms of Hp, Western blot and mass spectrometry after immunoprecipitation were carried out.

Materials and methods

Representative serum and liver samples from healthy and diseased cows (each n=3) as well as tissues from subcutaneous tail head and mesenteric fat were investigated. Cows were classified as diseased if they had clinical signs, the individuals studied herein had mastitis or metritis. Analyses were performed using an in-house polyclonal antibody raised in rabbits against bovine Hp. Hp serum concentrations were determined by ELISA as described previously (9). Using immunohistochemistry on paraffin embedded sections (10 μm), positive signals were obtained with the Hp antibody on adipose tissue after blocking endogenous peroxidase activity and unspecific binding sites on deparaffinized, rehydrated sections, that were thereafter incubated with the primary Hp antibody (1:2,000) and a secondary goat-anti-rabbit antibody (labeled with peroxidase; 1:200); staining was done with 3-amino-9-ethylcarbazol. To evaluate the different molecular weight forms of the Hp protein Western blot analysis was done with pooled sera and liver samples; mass spectrometry from individual serum and liver samples after immunoprecipitation of Hp was performed. For Western blot analyses, sample proteins (10-15 μg/ml) were separated by SDS gel electrophoresis after denaturation by dithiothreitol (DTT) and blotted on a hydrophobic polyvinylidene difluoride (PVDF) membrane. Proteins were identified by the primary Hp antibody (1:25,000) and a secondary goat-anti-rabbit antibody (peroxidase labeled; 1:50,000). Specific molecular weight bands were detected by enhanced chemiluminescence. For immunoprecipitation antibodies were covalently crosslinked to Sephadex G-15 and incubated overnight with 50 μg or 20 μl of proteins either from liver or serum samples, respectively. After immunoprecipitation pellets were resolved in HPLC-grade water and subjected to mass spectrometric analyses using MALDI-TOF-MS (Autoflex Speed, Bruker, Daltonics, Bremen, Germany) in linear mode with 2`,5`-dihydroxy acetophenone as matrix.

Results

Serum Hp concentrations mirrored the health status of the cows integrated into the present study. Mean Hp concentrations were 0.023 ng/ml and 3.145 ng/ml for healthy and diseased animals, respectively. Using the Hp antiserum in immunohistochemical analysis of adipose tissue sections, positive Hp staining signals were limited to adipocytes, however the staining intensity and the number of positive cells varied between the different fat depots (Figure 1).

To specify and identify this positive reaction, protein of fat homogenates was separated and analysed by Western blot. Positive signals were obtained at 25 kDa and at about 50 kDa (Figure 2, lanes 3 & 4). Western blot results from adipose tissue were compared with serum and liver samples. In addition to the positive signals at 25 kDa and 50 kDa, bands at about 20 kDa and 30 kDa were detectable (Figure 2, lanes 2, 5 & 6).

To identify the different Western blot bands, serum and liver samples from healthy and diseased animals were additionally analyzed by MALDI-TOF-MS. Some signals remain unidentified;

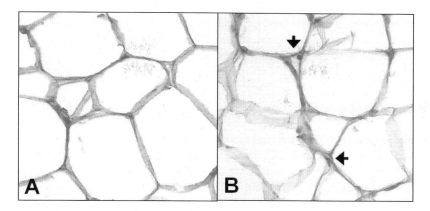

Figure 1. Positive reactivity of the Hp-antiserum in representative adipose tissue sections from visceral and subcutaneous fat depots detected by immunohistochemistry. A: mesenteric fat depot, B: subcutaneous tail head fat depot. Representative positive cells appear as red staining, arrows indicate examples of positive stainings.

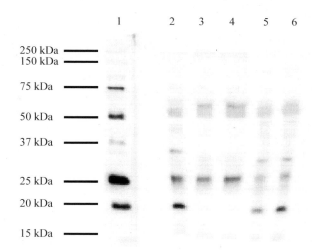

Figure 2. Western blot after reducing SDS-PAGE of Haptoglobin in bovine serum sample pool (lane 2), subcutaneous fat depot pool (lane 3: 10 µg protein & lane 4: 15 µg protein) and liver sample pools (lane 5: 10 µg protein & lane 6: 15 µg protein). Lane 1: molecular weight marker.

however, a peak of about 27.6 kDa was observed in all serum samples and was identified as the Hp-β chain. Interestingly, the molecular weight was around 350 Da higher as expected in all serum samples as well as in the applied standard. The Hp-α chain was not detected in serum by mass spectrometry. In liver samples, the 27.6 kDa peak was not detectable, but a peak at around 15.6 kDa which presumably represents the Hp-α chain, was evident. However, the amount of Hp represented by the size of the peaks was not as yet quantified. A difference of the isoforms could not be assigned to healthy and diseased cows. In all samples, independently from healthy status, hemoglobin α- and β-chains were detectable with a mass of 15.05 kDa and 15.95 kDa, respectively, for both Western blot analyses as well as for mass spectrometry.

Discussion and conclusions

The Western blot bands occurring at about 30 kDa and 20 kDa correspond to the heavy (β) and light (α) chain of Hp as demonstrated earlier in bovine serum (10). Differences between the molecular weight given by Western blot and mass spectrometry might be due to a shift of the molecular weight marker used for Western blot. The higher molecular weight of around 350 Da in serum samples might indicate a glycosylated form of the Hp-β chain. Although we could not sequence the ligands crossreacting with the Hp-antiserum at the moment, we suppose that different isoforms and modifications of the Hp-protein exist for different tissues. These isoforms could define different biological functions of extrahepatic Hp. Therefore the functional role of tissue-specific differences in Hp glycosylation patterns in dairy cattle remains to be elucidated.

References

1. AlDawood et al. 2010; Proc. 14th Int. Conf. Production Diseases in Farm Animals, Gent, Belgium, P106, ISBN: 978-5864-226-4.
2. Saremi et al. 2010; Proc. Of the 61st Ann. Meeting of the European Assoc. for Animal Production, Heraklion, Greece, P12, ISBN 978-90-8686-152-1.
3. Fain et al. 2004; J. Lipid Res. 45: 536.
4. Jensen 2006; Nature Rev. 7: 391.
5. Smeets et al. 2003; Cardiovascular Research 58: 689.
6. Cheng et al. 2007; Clin. Biochem. 40: 1045.
7. Fujimura et al. 2008; Int. J. Cancer 122: 39.
8. Turner 1995; Adv. Exper. Med. Biol. 376: 231.
9. Hiss et al. 2009; J. Dairy Sci. 92: 4439.
10. Hiss et al. J. Dairy Sci. 87: 3778.

Rapid protein production pipeline in advanced inducible *Leishmania tarentolae* expression system

S. Hresko[1], P. Mlynarcik[1], L. Pulzova[1,2], E. Bencurova[1], R. Mucha[2], T. Csank[1], M. Madar[1,2], M. Cepkova[1] and M. Bhide[1,2]
[1]*Laboratory of Biomedical Microbiology and Immunology, Department of microbiology and immunology, University of Veterinary Medicine and Pharmacy, Kosice, Slovakia; hresko@uvm.sk*
[2]*Institute of Neuroimmunology of Slovak Academy of Sciences, Bratislava, Slovakia*

Objectives

Leishmania tarentolae is a trypanosomatid protozoan host isolated from Moorish gecko *Tarentola mauritanica* [1], non-pathogenic to mammals, and can be used as a protein expression system. Proteins expressed by *Leishmania tarentolae* possess post-translation modifications close to mammalian-type, such as glycosylation, phosphorylation or prenylation. Moreover, these protozoa need similar handling as bacterial expression systems.

The aim of the study was to establish a rapid pipeline for expression and purification of recombinant proteins from *Leishmania tarentolae* expression system.

Material and methods

Expression vector pLEXSY_I-blecherry3 (Jena Bioscience, Germany) was modified as depicted in Figure 1. The first, GFP fusion tag was inserted into the 3×-terminus of open-reading-frame

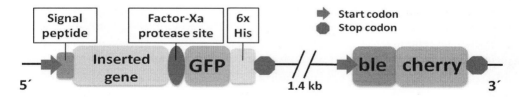

Figure 1. LBMI-LEXSY-GFP vector.
Vector contains two major expression cassettes: 1. for expression of target protein, 2. co-expression of bleomycin resistance marker fused with cherry fluorescence protein to monitor protein expression in the day light). In many cases co-expression of mCherry protein does not assure expression of protein of interest in the first expression cassette. To overcome this problem GFP fusion tag was inserted in the first cassette. Signal peptide ensures extracellular transport of overexpressed protein; factor-Xa site can be used to get only protein of interest, cleaved from rest of the GFP-polyhistidine tag, for downstream applications like crystallography.

(ORF) in the expression cassette. The second, to allow easy purification of overexpressed protein, polyhistidine tag was incorporated between GFP and stop codon. The third, factor-Xa protease site was inserted between protein of interest (ORF) and GFP tag. Modified plasmid was named as LBMI-LEXSY-GFP. Other features of this plasmid are discussed in the legend of Figure 1.

LBMI-LEXY-GFP vector was tested for overexpression of TNFR-Cys2 domain of CD40. In short, PCR amplified DNA of CD40 TNFR-Cys2 was ligated into LBMI-LEXSY-GFP and transformed into *E. coli* DH-5-α. Transformed colonies were selected on LB agar plates containing 50 ug/ml carbenicillin. Plasmid DNA was isolated from overnight culture (16 hours at 30 °C at 250 rpm shaking) of the single clone containing recombinant plasmid with the help of GenElute™ Plasmid Midi-Prep Kit (Sigma-Aldrich, Germany). Plasmid was linearized with SwaI enzyme digestion (1 hour at 37 °C) for homologous recombination of plasmid DNA into *Leishmania* chromosome (linearized fragment is presented in Figure 1). Linearized vector was electrophoretically separated on a 0.7% agarose gel and extracted using QIAquick Gel Extraction Kit (QIAGEN, Germany).

The LEXSY T7-TR strain (Jena Biosciences, Germany) of *Leishmania tarentolae* was cultivated as a static suspension in the dark at 26 °C in BHI medium supplemented with hemin at final concentration 5 µg/ml (essential for *Leishmania*), nourseothricin and hygromycin at final concentration 100 µg/ml each (for maintaining T7 polymerase and TET repressor genes in the host genome), and with penicillin and streptomycin (to prevent bacterial infections). Cells in mid-log phase were centrifuged for 3 min at 2,000xg at room temperature and a half of the medium was removed and cells were resuspended in remaining medium to get a suspension 10^8 cells/ml. Cells were chilled on wet ice for 10 min and mixed with linearized DNA. Electroporation was performed using Gene Pulser MXcell™ (Bio Rad, USA) set at 450 V and 450 µF in pre-chilled 2 mm gap electroporation cuvette resulting in a ~6 ms long pulse. Cuvette was put back on ice for 10 min and subsequently the cells were transferred to a fresh BHI medium with antibiotics and cultivated in the dark at 26 °C overnight. To select recombinant *Leishmania*, bleomycin at final concentration 100 µg/ml was added and incubation was continued at 26 °C for another 5 days. Recombinant *Leishmania* were passaged further and expression of TNFR-Cys2 GFP fused protein was induced with tetracycline at the final concentration 10 µg/ml. The expression of GFP fused protein and co-expressed mCherry protein was checked under microscope at 488 nm and 590 nm, respectively, at 24 and 48 hours after induction (Figure 2). After full protein expression (48 hours) cells were centrifuged for 3 min at 2,000xg at room temperature. The supernatant was collected and used for direct protein purification. In short, 20 µl of Talon metal affinity resin (Clontech) was washed twice with native wash buffer (50 mM sodium phosphate, 300 mM sodium chloride, 20 mM imidazole). To the resin 100 µl of BHI medium supernatant obtained from overexpression of protein and 900 µl of wash buffer was added. Binding of tagged protein to the resin was allowed for 10 min at 4 °C. Resin was then centrifuged at 10,000xg for 1 min and washed two times with wash buffer. Small amount (1 µl) of affinity resin was placed on glass slide and presence of GFP fused

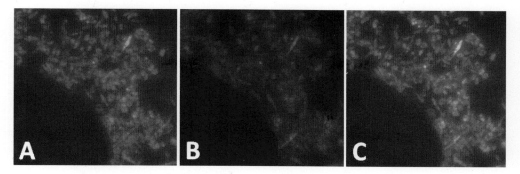

Figure 2. Fluorescence detection of expressed proteins in L. tarentolae.
A) green fluorescence of the fused GFP protein at 488 nm; B) red fluorescence of the co-expressed mCherry protein at 590 nm; C) merge of GFP and cherry fluorescence.

protein bound on agarose beads was checked under microscope at 488 nm (Figure 3). Protein bound on agarose beads was eluted with 20 µl of elution buffer (50 mM sodium phosphate, 300 mM sodium chloride, 200 mM imidazole). Elutions were desalted with ZipTip$_{C4}$ pipette tips (Millipore, MA, USA) according to manufacturer's instructions and protein was identified with MALDI-TOF analysis (Microflex, Bruker-Daltonics, Germany) by mixing 1 µl of elute with 1 µl of sDHB matrix. Mass spectra were recorded in the linear, positive mode at a laser frequency of 60 Hz (250 shots total, Figure 4). The spectra were calibrated using the protein Standard II (20,000 to 70,000 kDa range) from Bruker-Daltonics.

Results and discussion

The successful transformation, selection and expression of target protein in *Leishmania* were confirmed directly in living cells as the green and red fluorescence under a fluorescent microscope (Figure 2). The presence of secreted GFP fused CD40 TNFR-Cys2 protein in

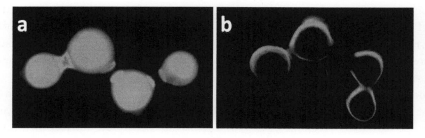

Figure 3. Detection of GFP fused protein from BHI medium.
Metal affinity agarose beads with GFP fused CD40 TNFR-Cys2 protein (a), agarose beads without GFP fusion protein (b).

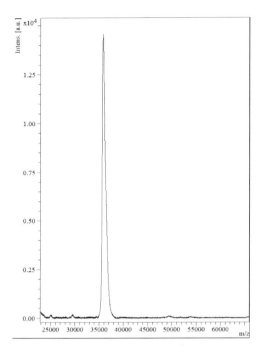

Figure 4. MALDI-TOF MS spectrum of purified GFP fused CD40 TNFR-Cys2 protein (~36 kDa).

cultivation medium was detected upon green fluorescence of silica beads (Figure 3). Finally, the result of MALDI-TOF analysis showed the present of a ~36 kDa protein (Figure 4), which corresponds to the mass of CD40 TNFR-Cys2 fused with GFP.

In summary, a rapid pipeline for overexpression of protein with LBMI-LEXY-GFP vector in *L. tarentolae* established in the present work has following advantages:
1. GFP fusion tag allowed easy and rapid assessment of the level of expression during the induction of target protein directly in living cells.
2. Polyhistidine tag served to capture and detect target protein, in its native state, directly from BHI medium (within 10 min), while GFP tag allowed detection of protein directly on agarose beads (less than a minute). Detection of target protein with classical methods like precipitation with TCA followed by SDS-PAGE separation and western blotting needs at least 2 days.
3. Capture of the tagged protein with metal affinity beads directly in BHI medium can be up-scaled to purify protein of interest in its native state. This is one of the major advantages against precipitation techniques, as many proteins may denaturate and lose their function. Proteins purified with metal affinity directly in BHI medium can be the fastest and safest method for downstream applications like in protein:protein interaction assays or study of co-enzyme activities.

4. Factor-Xa cleavage site allows separation of protein of interest from GFP and His tags. Such cleaved proteins are the most suitable for downstream applications like crystallography.

Acknowledgements

Work was performed in collaboration with Dr. E. Chakurkar, at ICAR Goa India, mainly for transfection and in vivo protein production. Financial support was from APVV-0036-10 and VEGA – 2/0121/11.

References

1. Wallbanks, K.R. *et al.* (1985): The identity of *Leishmania tarentolae* Wenyon 1921. Parasitology 90, 67-78.

Brucellosis and proteomics: an approach in Albania

Jani Mavromati
Veterinary Medicine Faculty, Agricultural University of Tirana, Albania;
j.mavromati@hotmail.com

Background

Brucellosis is a zoonotic infectious disease with adverse effects on both public and animal health (Corbel, 1997). The infection is present in many areas of the world, particularly in some Mediterranean (Kirandijski, 2009), Middle Eastern countries as well as in Albania. (Hajno *et al.*, 2010).

Purpose of the study

The main purpose of the study is the application of proteomics for the monitoring of vaccination for the prevention of Brucellosis on sheep.

Expected achievements and benefits

The expected result of this study is the understanding of the serum proteomes differences between sheep free of *B. melitensis* and sheep vaccinated with REV.1 vaccine.

Proteomics is increasingly at the forefront of biomedical investigation (Doherty *et al.*, 2008). Eschenbrenner *et al.*, 2010, investigated and compared the proteome of laboratory-grown strain Rev 1 and 16M, but no information about sheep serum proteomic after vaccination is available. Proteomics can help us in the process of monitoring the quality of vaccination of animals for the successful eradication of the Brucellosis infection in Albania.

Materials and methods

The study was conducted on 17 sera received on 20 October 2011 from the jugular vein from a flock of sheep in the Southern Albania free of brucellosis. 22 other sera were received on 22 October 2011 from a flock of sheep (Replacements) from a herd vaccinated with Rev. 1 on 22 July 2011, through the conjunctival route. In both flocks the Rosa Bengal and ELISAs tests were applied to reassure that the first flock was free of brucellosis, as well as that the second flock was vaccinated. Sera were transported with conformity rules and were kept at a temperature of -20 °C until the initiation of the analysis. The tests were conducted in Mass spectrometry Laboratory–GIGA – Proteomics, University of Liege, Belgium during November 2011. After the preparation of the samples in the laboratory, a differential proteomic study using a label free LC-MSe method was conducted (nano-2D UPLC coupled to a Synapt™ HDMS™ MSe, Waters; Figure 1).

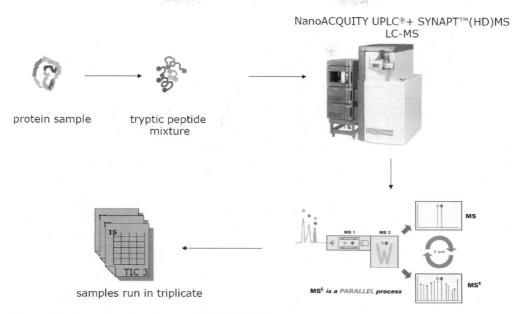

Figure 1. Schematic view of synapt™ HDMS™ G1 mass spectrometer.

Table 1. Number of identification proteins using sheep NCBI data base and mammalian data base.

Categories	Identification of the protein's number in the Sheep NCBI data base	Identification of the protein's number in the reviewed mammalian data base
Sheep free (3 pools)		
SFP1	47 proteins	114 proteins
SFP2	45 proteins	105 proteins
SFP3	49 proteins	128 proteins
Sheep vaccinated (3 pools)		
SVP1	51 proteins	109 proteins
SVP2	55 proteins	102 proteins
SVP3	53 proteins	101 proteins

Results

In Table 1, the numbers of identified proteins using Sheep NCBI data base and reviewed mammalian data base are listed. The number of identified proteins is about 50 proteins and 110 proteins respectively in each sample pool.

Discussion

After the filtration of the data from the 3 replications of each group we have found about 628 proteins isoforms, that were taxonomied from different animals, not only from sheep. This preliminary study allows us to underline some issues for the proteomic analysis of sheep sera. Basically the two major difficulties were coming from the biological matrix that was used.

First, serum is known to be a very complex protein mixture having in addition a variety of high abundant proteins. The number of protein identified and quantified can be drastically increased by applying depletion/fractionation or protein quantity normalization method.

Second, the sheep protein sequences database was not yet completed. To circumvent this issue we used a homology protein identification approach by searching against a Mammalian database.

The results, especially by using the homology identification method against the Mammalian database, shows interesting findings. In other words, protein isoforms were identified from other taxonomies indicating that these sequence protein isoforms are not yet available in the NCBI sheep database but most probably present in the sheep serum. The preliminary interpretation of the analysis underlines two processes potentially involved in the vaccination process.
1. The first one is related to the peroxisome activity which seems to have been modified between the two sheep populations. This was shown by the fact that numbers of peroxisomes proteins partners are found to be overexpressed only in the group of vaccinated sheep. This was already discussed in the theory by Garcia-Bates *et al.* (2009).
2. The second interesting group of modulated proteins is related to amine oxidase activity.

These results need of course to be further investigated.

Acknowledgements

I would like to thank COST FA1002 for making this STSM possible. I would also like to thank Dr. Gabriel Mazzuckelli on Mass spectrometry laboratory–GIGA-Proteomics, University of Liege, Belgium for the analyses of sera.

Farm animal proteomics

References

Corbel MJ, 1997, Brucellosis: an overview. Emerg Infect Dis. 1997 Apr-Jun;3(2):213-21.

Doherty MK, and al, Proteomics and naturally occurring animal diseases: Opportunities for animal and human medicine. Proteomics Clin Appl. 2008 Feb;2(2):135-41

Eschenbrenner M, *et al*. Comparative proteome analysis of *Brucella melitensis* vaccine strain Rev 1 and a virulent strain, 16M. Journal of Veterinary Diagnostic Investigation Vol. 22 Issue 4, 524-530, Copyright © 2010

Garcia-Bates TM, and al, Peroxisome proliferator-activated receptor gamma ligands enhance human B cell antibody production and differentiation, J Immunol. 2009 Dec 1;183(11):6903-12. Epub 2009 Nov 13).

Hajno L *et al*, Small Ruminant Production and Brucelosis infection consecutiveness. The act of the XVIII Intenational Congeress of Mediterranean Federation of Health and Production of Ruminants, Durres, 2010.

Kirandijski T. Brucellosis in small ruminants in the Republic of Macedonia. Book of abstract, Meta net project Thematic Scientific Conference: Brucellosis in the Mediterranean Region, Struga-Ohrid, 2009.

Saliva proteomics: tool for novel diagnosis for farm animal diseases from body fluids

Jacob Kuruvilla[1] and Susana Cristobal[1,2]
[1]Department of Clinical and Experimental Medicine, Cell biology, Faculty of Health Science Linköping University, Linköping, Sweden
[2]IKERBASQUE, Basque Foundation for Science, Bilbao, Spain, Department of Physiology. Basque Country Medical, Bilbao, Spain; susana.cristobal@liu.se

The *Millennium Development Goals* are the increase of food availability in order to reduce hunger and to eliminate problems caused by unbalanced diets, which are most frequently devoid of protein. Maintaining animal health during production increases agricultural productivity enhances food security and leads to the highest quality and safety of food. It is therefore essential to support research aimed at improving the health and welfare of animals farmed for food production and to promote the sustainability of farm animal industries. The specific aim of this project is to develop novel diagnosis for farm animal diseases. There is an increasing awareness of the potential of proteomic technologies to study production animals however, the use of proteomic strategies to investigate animal health and disease has been limited by the lack of experience with animal body fluids and tissues, and identification of proteins from non-sequenced genomes.

By this study, we aim to explore and gauge differentially expressed proteins in saliva proteome that could serve as potential biomarkers for human or farm animal diseases. This methodology would be used to develop diagnostic tools which can be exploited in non-sterile conditions especially in farm animal housing and obtain preliminary diagnosis in situ previous to any veterinary visit or any transport a sample to a the veterinary clinic.

The salivary proteins and their expression pattern have been mapped with reference to circadian rhythm and gender. So a work-flow has been designed where males and female saliva samples could be compared and quantified using iTRAQ (Isobaric tag for relative and absolute quantification). However, just like other body fluids, saliva is also complex thereby detection of peptide biomarkers typically present at low concentrations is hampered by the 'masking' effect caused by a number of highly abundant proteins. We initially focused in the sample preparation with the aim to specifically remove high abundant proteins. We tested several strategies: ultrafiltration, combinatorial library of hexapeptides bound to beads to decrease high-abundance proteins and enrich low-abundance, and affinity chromatography for removal of amylase and albumin. Finally, nanoparticles functionalized in the core with different chemically molecular baits showing preferential high affinities for specific low-abundance proteins.

There are several strategies for eradicating this bottleneck during biomarker discovery, one of which is depletion of the abundant proteins like amylase, albumin and immunoglobulins. We have made use of a starch affinity column for removal of amylase and an albumin-IgG immune-affinity depletion kit and we can clearly see the increase in number of visible bands (Figure 1).

Another strategy is to fish out those low abundant proteins using using a porous, buoyant, core-shell hydrogel nanoparticles functionalized with very specific amino-containing dyes just as it is better to catch the fish from the sea rather than to drain water to catch them[1]. After depletion, the sample proteins are trypsin digested, iTRAQ labeled, fractionated on basis of their hydrophobicity (Figure 2) and run on a nano LC ion-trap mass spectrometer to identify and catalogue the protein/s of interest.

In the last years, the saliva proteome has been started to be investigated as a novel source of protein biomarkers. However, saliva proteomics has scarcely been explored in farm animal. The parotid saliva proteome of sheep and goat have been analyzed by MALDI-TOF and LC-MS/MS and over 100 proteins were identified. It highlights the potential of proteomics to give insights in the intake behavior of herbivores [2]. Recently, a first high-throughput proteomic analysis of parotid saliva has described age specific variation in protein expression associated with host defense proteins [3]. In the case of acute phase proteins as biomarkers of inflammatory conditions of domestic animals, a critical mass of knowledge has accumulated over recent years, so that there is now sufficient understanding of the pathophysiology of the response to support the use of these compounds as diagnostic tools in clinical settings [4]. The actual diagnostic tools for farm animal diseases are based on the utilization of expensive, complex instruments in labs such hematology analyzer, chemistry analyzer, immuno-chemistry analyzer and flow cytometer. It could take up to several days to get conclusive results. The development of novel diagnostic tools should be an on-farm tool simple, affordable, accurate, and sensitive

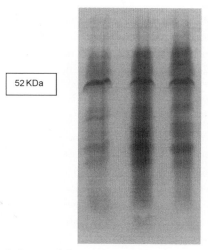

52 KDa

Figure 1. 50µg of protein loaded in each lane of an 18cm SDS PAGE. Lane 1: raw saliva. Lane 2: amylase depleted sample. Lane 3: albumin, amylase and IgG depleted sample.

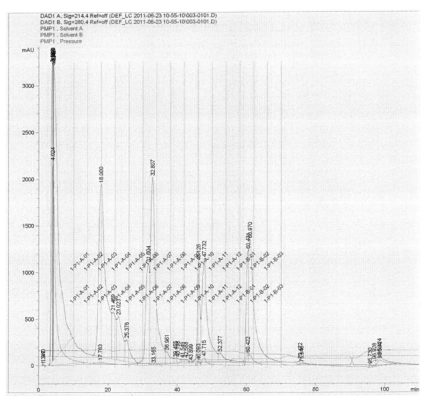

Figure 2. HPLC Chromatogram of 10 µg saliva proteins using solvents water and acetonitrile.

enough for the detection of subclinical level of infection that could affect to production. The state of art of proteomic technology could play an important role to discover the proteins or peptides that drastically change in expression after infection, and to apply this knowledge for the development of a simple micro-electro-mechanical tool.

References

1. Tamburro D, Fredolini C, Espina V, Douglas TA, Ranganathan A, Ilag L, Zhou W, Russo P, Espina BH, Muto G *et al*: Multifunctional core-shell nanoparticles: discovery of previously invisible biomarkers. J Am Chem Soc 2011, 133(47):19178-19188.
2. Lamy E, da Costa G, Santos R, Capela ESF, Potes J, Pereira A, Coelho AV, Sales Baptista E: Sheep and goat saliva proteome analysis: a useful tool for ingestive behavior research? Physiol Behav 2009, 98(4):393-401.
3. Ambatipudi KS, Lu B, Hagen FK, Melvin JE, Yates JR: Quantitative analysis of age specific variation in the abundance of human female parotid salivary proteins. J Proteome Res 2009, 8(11):5093-5102.
4. Eckersall PD, Bell R: Acute phase proteins: Biomarkers of infection and inflammation in veterinary medicine. Vet J, 185(1):23-27.

Proteomics in biomarker detection and monitoring of pancreas disease (PD) in atlantic salmon (*Salmo salar*)

Mark Braceland[1], Ralph Bickerdike[2], David Cockerill[3], Richard Burchmore[1], William Weir[1] and David Eckersall[1]
[1]*University of Glasgow, Glasgow, United Kingdom; m.braceland.1@research.gla.ac.uk*
[2]*BioMar Ltd., Grangemouth, United Kingdom*
[3]*Marine Harvest (Scotland) Ltd., Newbridge Midlothian, United Kingdom*

Introduction

The species of Salmonid Alphavirus (SAV) consists of six closely related subtypes and belongs to the genus *Alphavirus* within the family *Togaviridae* (McLouglin and Graham, 2007). SAV's are the aetiological agent of pancreas disease (PD) and sleeping disease (SD) of farmed Atlantic salmon, *Salmo salar*, and rainbow trout, *Oncorhynchus mykiss*. These diseases cause significant loss to aquaculture production in Europe. PD is characterized by lethargy and other behavioural modifications, sequential acute necrosis of the pancreatic acinar cells, cardiomyopathy, and skeletal muscle fibrosis and degeneration. Since the characterisation of PD a number of diagnostic tools have been developed such as; histopathology, RT-PCR SAV detection, and pathogen-specific antibody detection. However, little focus has been placed on investigating the processes involved in the return to homeostasis from SAV infection/ PD and how biological indicators of disease could be used as non-destructive tools for its identification and monitoring of disease progress. Therefore, this study investigates the modification of the serum proteome profile caused by PD, using SAV3 as the aetiological agent, in order to identify serum biomarkers of PD.

Materials and methods

Seven hundred Atlantic salmon (*Salmo salar* L.) parr of mean weight 30 g (<15%CV) were randomly distributed into 1 m^3 tanks and allowed to feed and grow for 42 days, after which 60 fish from each tank were bulk weighed and transferred into twelve 0.6 m^3 tanks. Additional fish from the tanks were maintained separately to be used as Trojan shedders of virus. These were allowed to acclimatise for 5 days and marked by clipping the adipose fin and injected with SAV 3 infected CHSE cell culture supernatant at ca. 105 TCID/fish into their intraperitoneal cavity. Inoculated Trojans were added to each of the twelve challenge tanks at 20% of the total tank population. Cohabitant fish were sampled at 0, 2, 3, 4, 5, 6, 8, 10 and 12 weeks post challenge (wpc). At each time point 9 fish per tank were killed by lethal overdose of anaesthetic (MS-222, Pharmaq) and blood collected in non-heparinised vaccutainers for preparation of serum. Fish sampled at time point 0 were sampled before adding Trojan shedders. For proteomic analysis one microlitre of serum from all serum samples from each tank was taken and one pool for each time point created. Protein concentrations of these were determined and pools diluted to an equal protein loading of 208 μg in rehydration buffer for 2-dimension electrophoresis analysis in duplicate (equipment & reagents from Biorad Ltd, Hemel Hempstead UK). Isoelectric focusing was carried out using 11cm immobilized pH gradient strips with a pH range of 3 to 10. Strips were then run on SDS-PAGE gels with XT MOPS running buffer, stained in Coomassie brilliant blue G-250 dye, de-stained and scanned for gel image analysis using Progenesis SameSpots 2D gel image analysis software (Nonlinear Dynamics Ltd, Newcastle, UK). Protein spots were identified which were differentially expressed through time. Initial results were filtered using the programme's statistical analysis function, with only those with a power value of >80% and ANNOVA significance score of <0.05 between replicate gels, being chosen for protein identification. These spots were excised manually by scalpel and subjected to in-gel trypsin digestion prior to identification via Bruker AmaZon ion trap mass-spectrometry and comparison to the MASCOT protein database. In addition, spot intensities throughout the time course, given by 'SameSpots', were analysed using ArrayStar software (DNAstar) using spot intensities to display expression through time and show the relationship between weeks post challenge by K-means un-centred Pearson, and to group proteins with similar expression profiles by Euclidean distance.

Results and discussion

A total of 72 proteins were established as being significantly differentially expressed (Figure 1) and were identified via MS/MS. (full list of identified spots can be seen in Figure 2).

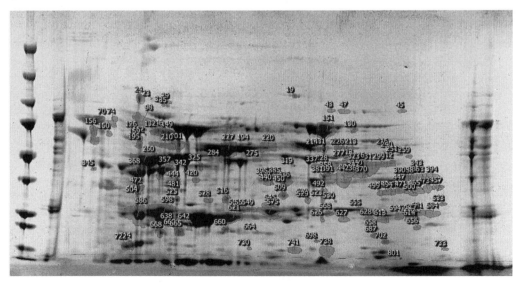

Figure 1. Image of 2D gel using serum from a week 0 pool with spots which were differentially expressed through time identified by numbers.

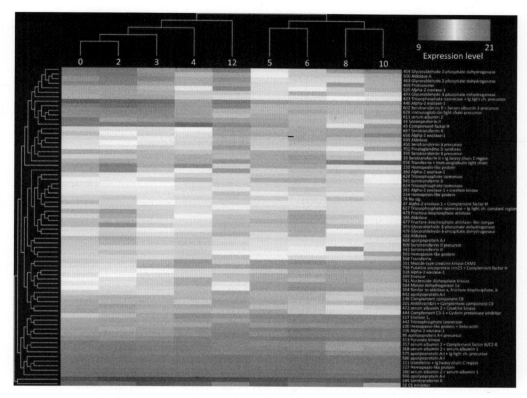

Figure 2. Arraystar heat map hierarchical clustering of spot intensities illustrating the relationship between weeks post challenge by K-means un-centred Pearson (top of diagram). The right-hand side gives spot number and corresponding identity from MS/MS. Grouped by profile similarity are on the left-hand side.

Among the proteins identified were a number of well-established biomarkers of viral diseases which in other species cause similar damage to tissues as in PD. These included creatine kinase, malate dehydrogenase, enolases, and aldolase. These biomarkers of tissue injury peak in spot intensity when pathological damage to tissue such as pancreas and muscle is at its worst during PD in weeks 6 to 8. In addition, a number of humoral immune system proteins were identified, for instance; complement components, hemopexin, and transferrin, whilst both light and heavy chains of immunoglobulin M were detected. Interestingly, a number of unexpected increases in protein were observed with glyceraldehyde 3-phosphate dehydrogenase (GAPDH) showing a rapid increase in intensity between wpc 4 and 6 and then rapidly falling to basal levels again by 10. A number of transferrin fragments were up regulated during PD whilst others were down regulated. K-Means un-centred Pearson of spot intensities demonstrates the similarity of the proteome of pools between wpc illustrating disease progression with large differences between pre and post wpc 4 until wpc 12 by which there is a return to levels similar to those pre 4wpc. In addition, Arraystar analysis of spot intensities also makes it possible to explore the relationship between proteins and enzymes as those which show similar expression patterns have been grouped together.

Conclusion

This study has highlighted a number of potential serum biomarkers for PD in Atlantic salmon which includes leaked enzymes from damaged tissues and humoral components of the innate and adaptive immune system. In addition, analysis of the expression profiles of proteins demonstrated the progression of disease from SAV infection to recovery, indicating that it may be possible, in a non-destructive manner, to estimate the disease stage of an individual based on its serum protein composition. Furthermore, there is clear grouping of enzymes which are biomarkers of tissue damage indicating the usefulness of proteomics in monitoring pathology. The observation that specific transferrin fragments were increased in abundance during PD of interest as it has been shown that in goldfish (*Carassius auratus*) transferrin fragments are not simply the result of post-secretion degradation of full length transferrin but are also secreted in a truncated form which can induce the nitric oxide (NO) response of macrophages (Stafford and Belosevic, 2003), with NO being both an antiviral-agent and immune system modulator (Akaike and Maeda, 2000). Furthermore, it has described that NO and GAPDH possess a strong affinity within cells and whilst no extracellular relationship between GAPDH and NO has been discovered it is interesting that transferrin fragments involved in NO macrophage activation and GAPDH both increase in concentration with GAPDH increasing and reaching its peak one week subsequent to the transferrin fragments.

Acknowledgments

BBSRC (Case), Biosciences KTN, Biomar Ltd and Marine Harvest Ltd are gratefully thanked for their support.

References

Akaike T, Maeda H (2000) In: *Nitric Oxide: Biology and Pathobiology.* Ignarro L J, editor. San Diego: Academic; pp. 733-745

McLoughlin MF, Graham DA (2007) *Alphavirus infections in salmonids- a review.* Journal of Fish Disease 30:511-531

Stafford JL, Belosevic M (2003). *Transferrin and the innate immune response of fish: identification of a novel mechanism of macrophage activation.* Developmental and Comparative Immunology, 27:539-54

Results of proteomic mapping into milk infected by *Streptococus agalactiae*

Monika Johansson[1], Amaya Albalat[2], Aileen Stirling[2] and William Mullen[2]
[1]*Department of Food Science, Swedish University of Agricultural Sciences, 75323 Uppsala, Sweden; monika.johansson@slu.se*
[2]*Biomarkers and Systems Medicine, University of Glasgow, Glasgow, Scotland, United Kingdom*

Streptococcus agalactiae is one of the most invasive mastitis pathogens in Sweden. It is considered one of the major causes of economic losses to dairy producers. Infections are spread from infected cows to non-infected cows during milking via milking machines, milkers' hands, and teat cleaning materials such as towels used on more than one cow. The infection results in subclinical or clinical mastitis, which decreases milk production and increases the somatic cell count (SCC) of the quarter. As with other causes of mastitis, *S. agalactiae* may cause heat, pain and swelling of the udder as well as abnormal milk consisting of white to yellow clots and flakes.

The bovine milk fraction consists for ± 90% w/w of α-lactoalbumin (α-LA), β-lactoglobulin (β-LG), α_{S2}-casein (α_{S2}-CN), α_{S1}-casein (α_{S1}-CN), κ-casein (κ-CN) and β-casein (β-CN); (Figure 1). The quality of dairy products, as texture, flavours and shelf life depend on these milk proteins. Hence, is not surprising that the activity of both endogenous and exogenous

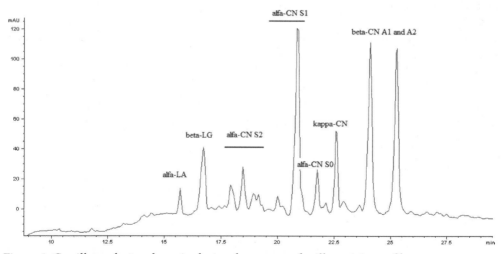

Figure 1. Capillary electrophoresis electropherogram of milk protein profile.

proteinases in milk have been studied over many years. However, very little is known about the effects of bacterial proteinases and their effect on proteolyses of whey proteins and caseins.

The extent of protein degradation caused by mastitis pathogen *S. agalactiae* at bovine milk was evaluated. In addition, the extent of protein degradation caused by different strains of *S. agalactiae* was also investigated. Pulsed-field gel electrophoreses (PFGE) was used as molecular typing tool for strain differentiation. The PFGE patterns for this bacteria showed great variation. Among the 96 isolates (collected from 43 cows on 12 farms) 11 distinct PFGE profiles were observed. Six strains were selected for evaluation by capillary electrophoreses for their ability to degrade the whey proteins and caseins.

In previous studies we showed that all six strains affected the relative concentration of milk proteins. However, the strains of *S. agalactiae* belonging to the same phylogenic group did not present the same patterns of degradation. As with most infections the most confirmative diagnosis is by bacterial culturing and testing in diagnostic laboratory. However, this can take a number of days to produce results and it would be an important aid to the dairy industry to have a rapid test that would allow early identification of this infection. Urinary proteomics has been used to identify biomarkers of a number of chronic diseases that can provide both diagnostic and prognostic information. Urine is an ideal biofluid candidate for proteomics due to its relatively simple composition and narrow dynamic range of concentrations and ease of accessibility.

As part of a join COST Action STMS, the Swedish University of Agricultural Sciences and the University of Glasgow exchanged knowledge on the preparation of bacterial strain of *S. agalactie* and urinary proteomic biomarker research. We have adapted the technology used in urinary proteomics for use in veterinary diagnostics of milk infected with *S. agalactia*. In the investigation carried out as part of the COST action STSM we used UHT milk that had been

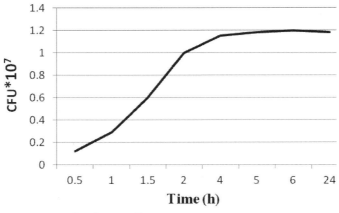

Figure 2. Streptococcus agalactie *growth curve.*

incubated with known strains of *S. agalactiae* and for known durations of incubation at 37 °C. Samples were collected at 0.5, 2 and 6 h after incubation and stored at -70 °C until extracted for proteomic analysis. All strains grew well, reaching a final concentration of approximately 10^7 CFU ml^{-1} after 6 hours (Figure. 2). Ten replicates of each sample plus additional time points for one sample were also produced. Each sample was put through a 20 kDa centrifugal filter until 1.1 ml of filtrate was obtained. The filtrate was desalted on a PD-10 column, lyophiliysed and stored a 4 °C until re-suspension to a final protein concentration of 0.7 μg/μl.

The samples were then analysed by CE-MS and the peptide profile and concentration of each sample measured. Replicates were compiled and compared between different controls and different strain and different time points.

The results of the analysis of these samples will be presented.

Acknowledgements

The work was funded by a COST action STMS to Monika Johansson and William Mullen.

Adhesion of *Francisella* to endothelial cells is also mediated by OmpA:ICAM-1 interaction

R. Mucha[1], E. Bencurova[2], M. Cepkova[2], P. Mlynarcik[2], M. Madar[1,2], L. Pulzova[1,2], S. Hresko[2] and M. Bhide[1,2]
[1]*Institute of Neuroimmunology of Slovak Academy of Sciences, Bratislava, Slovakia;* rastislavmucha@gmail.com
[2]*University of veterinary medicine and pharmacy in Kosice, Kosice, Slovakia*

Objectives

Tularemia (rabbit fever) is a serious infectious disease caused by *Francisella tularensis* infecting human and farm animals including cattle, rabbits, horses and sheep. The disease is economically important in rabbit industry. It is already known that *Francisella* readily adhere to various cells like macrophages, epithelial cells, endothelial cells, etc. to evoke self-internalization or crossing of various cell barriers [1-3]. There is still a lack of data that reveals interactions between pathogen ligands and host receptors expressed on the endothelial cells, which are crucial in initial steps of infection. In our previous study we showed that ICAM-1 is the most probable adhesive molecule for *Francisella* on endothelial cells (EC). The objective of this study was to identify protein candidate(s) from *Francisella* interacting with ICAM-1.

Material and methods

F. tularensis subsp. *holarctica* LVS (F. LVS) was cultured on enriched chocolate agar containing 1% glucose and 0.1% L-cystein. Whole cell lysate was prepared by sonication on ice and the total protein concentration was measured by the Bradford method.

To preprare recombinant ICAM-1, mRNA from the rat liver was isolated and reverse transcripted into cDNA. Whole protein coding region of ICAM-1 was amplified by PCR, ligated in to pYEBME1 plasmid (patent pending number 00089-2011, Slovak patent agency) and transfected in to *Saccharomyces cerevisiae* YPH501 strain. Expression of rICAM-1 protein was induced by galactose in uracil dropout medium. Cells were sonicated and His-tagged ICAM-1 protein was purified with Talon resin (Clontech, USA).

For identification of interacting protein from pathogen His-tagged ICAM-1 was imobilized on metal affinity beads (MB-IMAC, Bruker Daltonics) and hybridized with whole cell lysate of *F. tularensis* subsp. *holarctica* LVS. After stringent washings protein complexes were eluted and fractionated on SDS-PAGE. Interacting partner of ICAM-1 observed on PAGE (~38 kDa) was excised, digested with trypsin and peptide mass fingerprint was done with MALDI-TOF and NCBInr blast search which gave maximum identity with OmpA of *Francisella*.

To confirm interaction between ICAM-1 and OmpA the nucleotide sequence coding outer membrane domain of OmpA was amplified and ligated into pQE-GFP plasmid to get chimeric GFP fused OmpA ~45 kDa. *E. coli* M15 were transformed and His-tagged recombinant OmpA was overexpressed.

Western blot was applied to detect interaction between recombinant OmpA and ICAM-1. EC proteins were fractionated by non-reducing SDS-PAGE and transferred onto nitrocellulose membrane, blocked with 2% BSA in TTBS (0.5% Triton X, 25 mM Tris, 150 mM NaCl, pH 7.2) and hybridized with recombinant His-tagged rOmpA. Bound His-tagged candidate was detected with Ni-HRP conjugate and ECL.

To confirm interaction between ICAM-1 and rOmpA found in WB, pull-down assay was performed wherein OmpA was bound on affinity beads (Talon resin, Clontech) and hybridized with whole cell lysate of EC overnight at 4 °C. Protein complex was eluted and subjected for MALDI mass spectrometry (Bruker-Daltonik) using sDHB saturated in TA50 solution as matrix.

Results

In our previous study we identified ICAM-1 from host EC as potentially important adhesion protein for F. LVS. To explore further interaction(s) between ICAM-1 and pathogen ligand(s), the recombinant form of ICAM-1 was designed and overexpressed. With the help of MB-IMAC we showed that ~38 kDa protein of F. LVS interacts with rICAM-1 (Figure 1). Peptide mass fingerprinting identified this protein of F. LVS as OmpA (Figure 2). To verify ICAM-1:OmpA interaction recombinant form of OmpA was overexpressed in *E. coli* and the interaction was confirmed by western blot (Figure 3) and pull down assay (Figure 4).

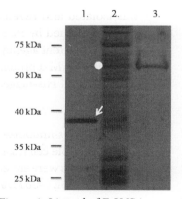

1 – Ligand of F.LVS co-precipitated with rICAM-1 in MB-IMAC (arrow). rICAM-1is showed by dot.

2 – whole cell lysate of F. LVS.

3 – only rICAM-1 captured on MB-IMAC beads and eluted (negative control).

Figure 1. Ligand of F. LVS interacting with rICAM-1 detected by modified MB-IMAC.

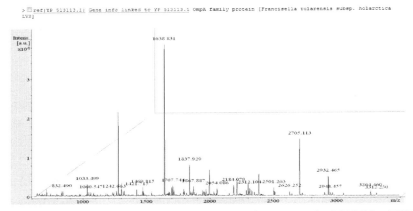

Figure 2. Identification of F. LVS ligand interacting with rICAM-1by MALDI – TOF analysis

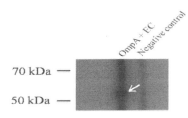

Proteins of EC were separated by SDS-PAGE, transferred on nitrocellulose membrane and hybridized with rOmpA protein. rOmpA bound on receptor from EC (arrow) was detected by Nickel HRP-ECL chemiluminiscence. As a negative control lysate of EC was incubated with TBS and subjeted for HRP-ECL detection.

Figure 3. Detection of interaction between rOmpA of F. LVS and receptor from EC.

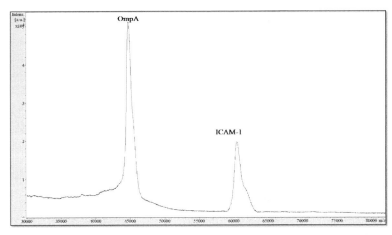

Recombinant protein OmpA was bound on the Talon beads. The beads were then incubated with whole cell lysate of rat EC and the protein complex was eluted and proteins were identified on MALDI-TOF.

Figure 4. Interaction between recombinant proteins rOmpA and rICAM-1.

Farm animal proteomics

Recombinant protein OmpA was bound on the Talon beads. The beads were then incubated with whole cell lysate of rat EC and the protein complex was eluted and proteins were identified on MALDI-TOF.

Conclusion

This work presents protein candidates most probably taking part in adhesion process of Francisella to endothelial cells. Adhesion process is crucial primary step of bacterial invasion process. It is also needed for internalization of bacteria in different immune cells like macrophages as well as adhesion processes necessary for translocation of bacteria across various barriers like epithelial or endothelial. This finding could help to understand molecular basis of pathogenesis of Tularemia.

Acknowledgements

We thank Dr. E.B. Chakurkar from Indian Council of Agricultural Research (ICAR) Complex (India) for his the excellent assistance in preparation of recombinant form of proteins. The work was supported by APVV-0036-10 and VEGA- 2/0121/11.

References

1. Kugeler KJ, *et al*: Isolation and characterization of a novel Francisella sp. from human cerebrospinal fluid and blood. J Clin Microbiol 2008, 46(7):2428-2431.
2. Abril C, *et al*: Rapid diagnosis and quantification of Francisella tularensis in organs of naturally infected common squirrel monkeys (Saimiri sciureus). Vet Microbiol 2008, 127(1-2):203-208.
3. Park CH, *et al*: Pathological and microbiological studies of Japanese Hare (Lepus brachyurus angustidens) naturally infected with Francisella tularensis subsp. holarctica. J Vet Med Sci 2009, 71(12):1629-1635.

Proteins overexpressed in ovine pulmonary adenocarcinoma

Anna Kycko and Michal Reichert
Department of Pathology, National Veterinary Research Institute, Pulawy, Poland;
reichert@piwet.pulawy.pl

Introduction

Ovine pulmonary adenocarcinoma (OPA) is a transmissible lung cancer of sheep caused by Jaagsiekte sheep retrovirus (JSRV). The virus induces neoplastic transformation of secretory epithelial cells of sheep lung: alveolar type II pneumocytes and Clara cells. The disease is present worldwide, apart from Australia and New Zealand and it has been eradicated from Iceland. In Poland, the suspicion of its occurrence was reported in 2002 (3). OPA has a long incubation period. It takes at least 3 weeks in experimentally infected lambs, but usually lasts over 2 years in adult sheep to the occurrence of clinical signs of respiratory tract distress. Rapid breathing reflects the extent of tumor development in the lungs. Raising the back and lowering the head of the sheep may cause frothy mucoid fluid accumulated within the respiratory tract to leak out from the nostrils (15). The infection is persistent and there is no immune response, due to the presence of similar endogenous retroviruses. At present there is no approved serological or other diagnostic blood test for OPA, especially before clinical signs occur, and only histological examination of the lungs and PCR based methods to detect the virus (usually when the disease is fully developed) is a way to diagnosis of OPA (6). OPA is clinically and histologically similar to human lung adenocarcinomas of mixed subtypes (5). The possibility of experimental induction of the tumor in animals makes it a good model for the study of human lung tumors.

The present study was intended to examine the proteomes of OPA affected and unaffected sheep lung tissues for the identification of proteins differentially expressed in ovine pulmonary adenocarcinoma affected lung.

Material and methods

The experimental group of sheep involved nine animals. Five lambs (1-3 weeks old) were inoculated intratracheally with JSRV clone (kindly provided by Professor M. Palmarini from the University of Glasgow, Scotland). The non-infected group of four animals served as negative control. Three infected animals died 1-4 weeks after the first visible signs of dyspnea had occurred. The remaining animals were euthanized after 1.5 year. Lung sections were collected during the necropsy for histopatological and immunohistochemical examination. The lung samples were also frozen at -70 °C for molecular biology and proteomic study.

OPA was confirmed histopathologically and using PCR. In one case of negative control, bronchopneumonia was revealed.

Frozen lung tissue sections from healthy and adenocarcinoma affected sheep were lysed in appropriate buffer. The two-dimensional (2D) electrophoresis of the protein lysates was performed, using 17 cm IPG Strips of pH gradient 3-10, each sample in two repetitions. The resulting gels were visualized using either silver or Coomassie Blue staining, then scanned and analysed using PDQuest software (BioRad, USA). The differentially expressed spots were excised and analysed with the LC- MS/MS method for the protein identification.

For the immunohistochemistry, the following commercial antibodies against cytokeratin 19 and aldolase A were used as the primary antibodies: ALDOA / Aldolase Goat anti-Rabbit Polyclonal (10 mg/ml, LS-C34940 Life Spam Biosciences, dilution 1:800), and Cytokeratin 19 (312-335) Mouse anti-Human Monoclonal (AM08421PU-N, Acris Antibodies, dilution 1:80). Immunohistochemical study was performed using labeled streptavidin-biotin visualization system – HRP/LSAB+ kit (Dako, Denmark).

Results

There were 14 spots presenting at least 2-fold higher expression in each neoplastic tissue sample as compared with non-neoplastic ones. These differential spots were excised and analysed with the mass spectrometry.

The analysis of the spots using mass spectrometry revealed, that in 11 out of 14 spots, there were more than one polypeptide detected, indicating overlapping of proteins. The single proteins that were identified in 3 spots were: cytokeratin 19, aldolase A and mangan superoxide dismutase. For immunohistochemical study we used antibodies against cytokeratin 19 and aldolase A as these proteins were characterised by over 3-fold overexpression in the neoplastic samples.

Immunohistochemistry: in all specimens, normal, ciliated bronchiole epithelium as well as individual epithelial cells (type II pneumocytes) showed strong and diffuse expression of cytokeratin 19, aldolase A. Immunohistochemical analysis of CK19 and aldolase A expression in a lung adenocarcinoma showed abundant cytoplasmic staining in all tumor cells. The positive reaction for the presence of CK 19 and aldolase A was also observed in macrophages present in neoplastic specimens.

Discussion

In our study, the two dimensional electrophoresis together with mass spectrometry and immunohistochemistry allowed for revealing, that the two proteins: cytokeratin 19 and aldolase A were overexpressed in lung tissues of sheep afected with OPA. The expression of these proteins was analysed previously by other groups, and there are several reports describing their over expression in cases of human cancer.

The first analysed protein – cytokeratin 19 – belongs to the group of acidic (type I) cytokeratins (CK) that are intermediate filament proteins forming the cytoskeleton. CK 19 is the smallest known type I cytokeratin and is not paired with a basic cytokeratin, unlike the rest of proteins of this group. CK 19 is expressed specifically in periderm, the layer that envelopes developing epidermis. Together with cytokeratins 8, 9 and 18 CK 19 is expressed in simple, columnar epithelia such as the respiratory epithelium of the bronchial tree (11). Increased expression of this protein was observed in lung cancer cells in humans using immunohistochemical methods. The particularly strong overexpression was noted in squamous cell carcinoma and adenocarcinoma (4, 8). It was also found to be expressed in human epithelial malignancies including neoplasms of the liver (16), colon, stomach, pancreas (14) biliary tract (10), and breast (1). The study on the expression of cytokeratin in human adenocarcinoma of the lung using two dimensional electrophoresis demonstrated the presence of three isoforms of cytokeratin 19 similar in molecular weight, but differing in isoelectric points, indicating potential modifications of the protein. Two isoforms of lower pI the were of the strongest expression. In the same study there was also observed positive correlation between poor prognosis for patients affected with this neoplasm and high expression of CK19 protein isoforms of lowest pI values (7).

Aldolase A (fructose-1,6-(bis)phosphate aldolase), which was identified in the second analysed spot is an enzyme belonging to the class of lyases and is involved in glycolysis and gluconeogenesis. In the cell it was found in the nucleus and cytoplasm, and was described to be involved in the organization of actin filaments and cytoskeleton formation (9) Aldolase A expression was found to be increased in various neoplasms in humans, such as lung, liver, gastric or colorectal cancer. The high level of aldolase in cancer tissues is considered to be associated with anaerobic glycolysis, which is generally enhanced in neoplastic tissues (2, 13)

Our study is to our knowledge the first report of proteomic analysis of OPA. We expect, that using narrow pH gradients in 2D electrophoresis may reveal more spots in gels representing single differentialy expressed proteins. The proteomic methods prove to be useful for analysis of the biological nature of the ovine pulmonary adenocarcinoma, finding similarities between OPA and human lung neoplasms on the protein level, as well as revealing possible markers for the disease.

Acknowledgments

This work was supported by Project No.: N N308 256335 funded by the Ministry of Science and Higher Education of Poland.

References

1. Alix-Panabières C., Vendrell J.P., Slijper M., Pellé O., Barbotte E., Mercier G., *et al*. Full-length cytokeratin-19 is released by human tumor cells: a potential role in metastatic progression of breast cancer. Breast Cancer Res. 2009;11(3):R39.

2. Asaka M., Kimura T., Meguro T., Kalo M., Kudo M., Miyazaki T., Alpert E. Alteration of aldolase isozymes in serum and tissues of patients with cancer and other diseases. J Clin Lab Anal. 1994, 8: 144-148

3. Anonim, World Organization for Animal Health, Handistatus II – Multiannual Animal Disease Status 2006.

4. Chyczewski L., Nikliński J., Chyczewska E., Laudański J., Furman M. Immunohistochemical analysis of tissue localization of cytokeratin 19 in lung cancer. Rocz. Akad Med Białyst.1997; 42: 162-172.

5. De las Heras, M., L. Gonzalez L., Sharp J. M. Pathology of ovine pulmonary adenocarcinoma. Curr. Top. Microbiol. Immunol. 2003, 275, 25-54.

6. De Las Heras M., Ortin A., Salvatori D., Perez de Villareal M., Cousens C., Miguel Ferrer L., Miguel Cebrian L., Garcia de Jalon J.A., Gonzalez L., Michael Sharp J.: A PCR technique for the detection of Jaagsiekte sheep retrovirus in the blood suitable for the screening of ovine pulmonary adenocarcinoma in field conditions. Res Vet Sci 2005, 79, 259-264.

7. Gharib T.G., Chen G., Wang H., Huang C.C., Prescott M.S., Shedden K., Misek D.E., Thomas D.G., Giordano T.J., Taylor J.M., Kardia S., Yee J., Orringer M.B., Hanash S., Beer D.G. Proteomic analysis of cytokeratin isoforms uncovers association with survival in lung adenocarcinoma. Neoplasia. 2002;4(5):440-8

8. Kosacka M,. Jankowska R. Prognostyczne znaczenie stopnia ekspresji cytokeratyny 19 w niedrobnokomórkowym raku płuca. Pneumonol Alergol Pol 2007; 75: 317-323

9. Kusakabe T., Motoki K., Hori K. Mode of interactions of human aldolase isozymes with cytoskeletons. Arch Biochem Biophys. 1997;344(1):184-93.

10. Lódi C., Szabó E., Holczbauer A., Batmunkh E., Szíjártó A., Kupcsulik P., *et al*. Claudin-4 differentiates biliary tract cancers from hepatocellular carcinomas. Mod. Pathol. 2006;19(3):460-9.

11. Moll R., Franke W.W., Schiller D.L., Geiger B., Krepler R. The catalog of human cytokeratins: Patterns of expression in normal epithelia, tumors and cultured cells. 1982 Cell 31: 11-24.

12. Naseem N., Reyaz N., Nagi A.H., Ashraf M., Sami W. Immunohistochemical Expression of Cytokeratin-19 in Non Small Cell Lung Carcinomas – An Experience from a Tertiary Care Hospital in Lahore International Journal of Pathology. 2010; 8(2): 54-59

13. Ojika T., Imaizumi M., Abe T., Kalo K. Immunochemical and immunohistochemical studies on three aldolase isozymes in human lung cancer. Cancer 1991;67: 2153-2158

14. Ohshio G., Imamura T., Okada N., Yamaki K., Suwa H., Imamura M., *et al*. Cytokeratin 19 fragment in serum and tissues of patients with pancreatic diseases. International Journal of Gastrointestinal Cancer. 1997;21(3):235-41.

15. Sharp J.M., DeMartini J.C.: Natural history of JSRV in sheep. Curr Top Microbiol Immunol 2003, 275, 55-79

16. Van Sprundel R.G., van den Ingh T.S., Desmet V.J., Katoonizadeh A., Penning L.C., Rothuizen J., *et al*. Keratin 19 marks poor differentiation and a more aggressive behaviour in canine and human hepatocellular tumours. Comp Hepatol. 2010 18;9:4.

Relevance of paraoxonase-1, platelet- activating factor acetylhydrolase and serum amyloid A in bovine mastitis

Romana Turk[1], Mislav Kovačić[2], Zlata Flegar Meštrić[3], Paola Roncada[4], Cristian Piras[5], Ante Svetina[1], Vesna Dobranić[6], Jelka Pleadin[7] and Marko Samardžija[8]
[1]Department of Pathophysiology, University of Zagreb, Zagreb, Croatia; rturk@vef.hr
[2]Pharmas d.o.o., Popovača, Croatia
[3]Institute of Clinical Chemistry, Clinical Hospital 'Merkur', Zagreb, Croatia
[4]Istituto Sperimentale Italiano L. Spallanzani, Milano, Sezione di Proteomica c/o Facoltà di Medicina Veterinaria, Università degli Studi di Milano, Milano, Italy
[5]University of Sassari, Department of Zootechnical Sciences, Faculty of Agrarian Sciences, Italy
[6]Department of Hygiene and Technology of Animal Origin and Food Staff, University of Zagreb, Zagreb, Croatia
[7]Laboratory for Analytical Chemistry, Croatian Veterinary Institute, Zagreb, Croatia
[8]Department of Reproduction and Clinic for Obstetrics Faculty of Veterinary Medicine, University of Zagreb, Zagreb, Croatia

Bovine mastitis is the most economically important production disease in dairy cows. The pathogenesis of mastitis involves an inflammatory reaction resulting from production of cytokines at the site of infection. Proinflammatory cytokines stimulate the synthesis of acute phase proteins (APPs) predominantly in liver but also in other tissues including mammary gland (Eckersall *et al.*, 2006). Serum amyloid A (SAA) is one of the major APP in ruminants whose concentration rises dramatically during acute phase response (APR). During APR, SAA associates rapidly with high density lipoproteins (HDL), displacing apo A-I (the major apolipoprotein of native HDL) and becoming the predominant apolipoprotein (apo SAA) in HDL. Native HDL contains paraoxonase-1 (PON1) and platelet-activating factor acetylhydrolase (PAF-AH), both antioxidative/anti-inflammatory enzymes being responsible for anti-inflammatory/anti-oxidative properties of HDL. Concomitantly with SAA enrichment of HDL during APR, PON1 and PAF-AH levels decrease within HDL, what impairs anti-oxidative properties of HDL which from an anti-inflammatory particle becomes a pro-inflammatory particle, i.e. acute phase HDL (AP-HDL) (Van Lenten at al., 1995).

In order to evaluate systemic inflammatory and oxidative stress response, we investigated serum amiloid A (SAA) concentration, PON1 and PAF-AH activities in serum of cows with subclinical and clinical mastitis.

The study was conducted on a total of 80 Holstein-Frisian dairy cows located on farms in the region of Eastern Croatia. All the cows underwent a physical examination and somatic cells count (SCC) and California mastitis test in milk samples were performed. Based on the results, the cows were divided into 3 groups. Group I (control, n=30) consisted of healthy cows with SCC below 200,000 cells/ml with negative California mastitis test and without any clinical sign of mastitis. Group II (subclinical mastitis, n=30) comprised cows without clinical signs of mastitis but with SCC above 200,000 cells/ml and a positive California mastitis test. Group III (clinical mastitis, n=20) consisted of cows with clinical signs of mastitis which include changes in milk appearance (flakes and clots in milk) and different stages of udder inflammation (hyperemia, edema, pain, udder enlargement and elevated udder temperature). Blood samples were taken from *v. coccygea* and centrifuged at 1,500 rpm for 15 min after clotting for two hours at room temperature. Serum samples were stored at -70 °C until analysis.

PON1 activity was assayed by the slightly modified method of hydrolysis of paraoxon previously described by Mackness *et al.* (1991) and Schiavon *et al.* (1996). PON1 activity was expressed in U/L as the amount of substrate hydrolyzed per minute and per liter of serum (μmolmin^{-1}/l). PAF-AH activity was determined by the spectrophotometric assay described by Kosaka *et al.* (2000). SAA concentration was assayed using the multispecies Tridelta Phase™ range ELISA SAA kit (Tridelta Development Ltd., Ireland).Triglyceride, total cholesterol and HDL-C concentrations were measured by standard commercial kits (Olympus Diagnostica GmbH, Hamburg, Germany and ThermoTrace, Victoria, Australia). All measurements were performed on automatic analyzer Olympus AU 600.

Subclinical and clinical mastitis results in significant changes in the activities of anti-inflamatory/antioxidative enzymes and acute phase protein serum amyloid A (Figure 1).

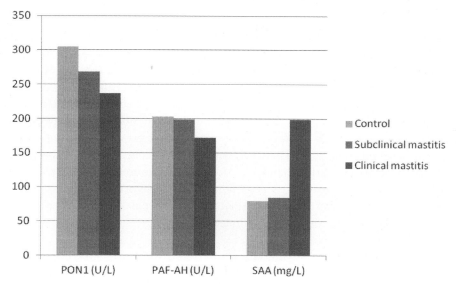

Figure 1. Serum PON1, PAF-AH and SAA in control cows and cows with subclinical and clinical mastitis.
[a,b]*Values with different superscript letters are significantly different.*

PON1 activity was significantly lower ($P<0.05$) in subclinical and clinical mastitis compared to control cows. PAF-AH activity was significantly lower ($P<0.05$) only in clinical mastitis but not in cows with subclinical mastitis. SAA concentration was statistically higher ($P<0.05$) in clinical mastitis and there were no significant difference between subclinical mastitis and healthy cows.

Triglyceride concentration was significantly higher ($P<0.05$) in the serum of cows with clinical mastitis, while total cholesterol and HDL-C concentrations were significantly lower ($P<0.05$) in the serum of those with clinical mastitis. There were no significant differences in concentrations in lipid status between healthy cows and those with subclinical mastitis (Table 1).

Table 1. Lipid status and calcium concentration (mean ± SD) in control cows and cows with subclinical and clinical mastitis.

Parameter	Control (n=30)	Subclinical mastitis (n=30)	Clinical mastitis (n=20)
Triglyceride (mmol/l)	0.11±0.02[a]	0.12±0.03[a]	0.14±0.03[b]
Total cholesterol (mmol/l)	7.9±1.5[a]	7.5±1.7[a]	4.9±1.5[b]
HDL-C (mmol/l)	4.2±0.8[a]	4.0±0.7[a]	2.9±0.7[b]

[a,b] Values with different superscript letters are significantly different.

The present study describes systemic acute phase response in cows with subclinical and clinical mastitis. Serum PON1, PAF-AH and SAA demonstrate different responses in subclinical and clinical mastitis. PON1 activity showed a significant decrease in subclinical and clinical mastitis compared to control while PAF-AH activity showed a significant decrease only in clinical mastitis. SAA showed a high response only in clinical mastitis being 2.5 fold higher in concentration than in healthy cows. SAA was not increased in subclinical mastitis indicating probably SAA expression only in the mammary gland. In the present study, only PON1 was lower in both subclinical and clinical mastitis compared to healthy cows, indicating that PON1 activity is a more sensitive marker then SAA. As PON1 is an antioxidant enzyme, low activity indicates an involvement of oxidative stress in the pathogenesis of mastitis. Since both PON1 and PAF-AH decrease within HDL during APR (Van Lenten *et al.*, 1995), both enzymes could be considered as negative acute phase reactants.

Our results showed that the inflammation interfered with lipid metabolism affecting concentration of total cholesterol, HDL-C and triglycerides (TG) in cows with clinical mastitis. HDL-C and total cholesterol were lower in clinical mastitis while TG concentration was higher in cows with clinical mastitis. The reasons for lipids alteration probably lie in the remodeling of lipoprotein particles during the interaction of APPs with plasma lipoproteins.

Hypertriglyceridemia during APR could contribute to triglyceride enrichment of HDL what could also contribute to decrease of PON1 and PAF-AH from HDL (Cabana *et al.*, 1999). A significant reduction in PON1 and PAF-AH might be a contributing factor to inflammation in the pathogenesis of subclinical and clinical mastitis.

The results demonstrate that SAA concentration in the serum is a reliable marker for diagnosis of clinical mastitis showing high response in cows with clinical form of disease but could not distinguish subclinical from clinical mastitis. On the other hand, lower PON1 activity in subclinical mastitis with moderate response demonstrates PON1 as a possible marker for diagnosis of subclinical form of disease.

References

Cabana V, Lukens J, Rice KS, Hawkins TJ, Getz G. HDL content and composition in acute phase response in three species: triglyceride enrichment of HDL a factor in its decrease. J Lipid Res. 1996;37:2662.

Eckersall PD, Young FJ, Nolan AM, Knight CH, McComb C, Waterston MM, *et al.* Acute phase proteins in bovine milk in an experimental model of Staphylococcus aureus subclinical mastitis. J Dairy Sci. 2006;89:1488-501.

Kosaka T, Yamaguchi M, Soda Y, Kishimoto T, Tago A, Toyosato M, *et al.* Spectrophotometric assay for serum platelet-activating factor acetylhydrolase activity. Clinica chimica acta. 2000;296:151-61.

Mackness MI, Harty D, Bhatnagar D, Winocour PH, Arrol S, Ishola M, *et al.* Serum paraoxonase activity in familial hypercholesterolaemia and insulin-dependent diabetes mellitus. Atherosclerosis. 1991;86:193-9.

Schiavon R, De Fanti E, Giavarina D, Biasioli S, Cavalcanti G, Guidi G. Serum paraoxonase activity is decreased in uremic patients. Clinica chimica acta. 1996;247:71-80.

Van Lenten B, Hama S, De Beer F, Stafforini D, McIntyre T, Prescott S, *et al.* Anti-inflammatory HDL becomes pro-inflammatory during the acute phase response. Loss of protective effect of HDL against LDL oxidation in aortic wall cell cocultures. Journal of Clinical Investigation. 1995;96:2758.

Proteomic characterization of serum amyloid A protein in different porcine body fluids

L. Soler[1,2], J.J. Cerón,[2] A. Gutiérrez[2], C. Lecchi[3], A. Scarafoni[4] and F. Ceciliani[3]
[1]Division of Livestock-Nutrition-Quality. Faculty of Bioscience Engineering. K.U. Leuven. Heverlee 3001, Belgium; laurasv@um.es
[2]Department of Animal Medicine and Surgery, Faculty of Veterinary Medicine. University of Murcia 30100 Espinardo, Spain
[3] Department of Animal Pathology, Hygiene and Veterinary Public Health. University of Milano 20133 Milano, Italy
[4]Department of AgriFood Molecular Sciences, University of Milano 20133 Milano, Italy

Introduction

Serum Amyloid A (SAA) proteins constitute a superfamily of small proteins (14-15 kDa) composed by a number of isoforms[1], of which SAA1/2 (pI 5-6) are the main circulating forms in most of the species, and SAA3 is the predominant SAA isoform produced extrahepatically (pI higher than 9)[1]. Although its structure has not been determined, human SAA is considered an apolipoprotein, and can be found in biological fluids as monomers, aggregated as different-order multimers or associated with other proteins, such as HDL or albumin[2,3]. Exact functions of SAA remain unclear, although immunomodulatory, antimicrobial and lipid metabolism-related functions have been described in the literature[1]. Recent reports have established a relationship between functionality and conformation, and so the exact function of SAA would depend on its aggregation state[4]. SAA is also involved in the pathogenesis of AA-amyloidosis, a disease that is rare in pigs[5]. It has been postulated that porcine resistance to AA amyloidosis could be a consequence of a singular SAA isoform pattern[5]. In fact, we recently described that the theoretical characteristics of the SAA cDNA sequences obtained from different pig tissues indicated a singular isoform pattern in this species, since the pig SAA main circulating form showed properties of local SAA[6]. To the author's knowledge, no studies have been performed to characterize pig SAA proteins by proteomics, probably due to the lack of pig-specific reagents. In the present study, the SAA conformation and isoforms were investigated in different porcine body fluids by 1-DE (SDS-PAGE) and 2-DE analysis and subsequent immunoblotting employing a specific in-house produced anti-pig SAA monoclonal antibody[6].

Material and methods

Analyzed fluids were porcine serum from diseased (SD), and healthy animals (SH), sow colostrum (COL), newborn piglet serum (SNP), porcine recombinant SAA (rSAA[6]) and purified haptoglobin[7] (Hp). All procedures involving animals were approved by the Murcia University Ethics Committee.

SDS-PAGE analysis

An amount of total protein of 15 µg (SD, SH, SNP and COL) or 5 µg (rSAA, Hp) of protein were run in 4-12% SDS-PAGE gels under reducing conditions according to standard protocols[8].

2-DE

First-dimensional immobilized pH-gradients (ReadyStrip IPG strips, 7 cm length, BioRad) were run under reducing and denaturing conditions in a *Multiphor* II Flatbed Electrophoresis System (GE) in linear gradients pH 3-10. An amount of total protein of 15 µg (silver staining) or 40 µg (immunodetection) was loaded. Second-dimensional electrophoresis was run on 10% or 12% acrylamide-bisacrylamide gels (80x73x1 mm) in a Mini-protean 3 electrophoresis cell (BioRad) according to standard protocols[8].

Immunodetection

Separated proteins were transferred onto PVDF membranes and SAA was identified with an in-house produced anti-pig SAA biotin-labelled mAb[6], followed by streptavidin-HRP conjugate (GE). For detection, either colorimetric staining (4-chloro-1-naphthol) or chemifluorescent signal detection (ECL Plus, GE) were used. For imaging with chemifluorescent detection, a Typhoon scanner was employed (GE). All the images were digitalized and analyzed by ImageQuant TL or ImageMaster 2D Platinum v7.0 image analysis software (GE). Lanes containing molecular weight markers were used as standards to calculate the molecular weight of unknown bands.

Results and discussion

Representative results from SDS-PAGE analysis is shown in Figs 1A (silverstainig) and 1B (immunodetection of SAA in same samples as in 1A). Immunoreactive bands of approx. 26, 40, 120 and 180 kDa were found in all serum samples, which can be attributed to different order SAA multimers and/or aggregates[3]. Approximately 15 kDa bands corresponding to SAA monomers could also be seen in SD2, SNP, COL and rSAA. In colostrum, an additional approx. 13 kDa band was found and no 27 kDa band could be seen. The apparent variability obtained might be attributed to the known influence of sample preservation and running conditions on SAA multimer conformation and aggregation[3]. The molecular weight of the observed immunoreactive bands matched with those of the reduced pig Hp[7], but no cross-reactivity of the anti-SAA antibody was observed with pig purified Hp (Figure 1B).

The analysis of rSAA by 2-DE (Figure 2A) confirmed the presence of different order multimers with different pIs: 15 kDa monomer with an approx. pI of 9.5 and 7.4, 42 kDa multimer with a pI of approx. 6.5 and 60 and 75 kDa, multimers with neutral pI (6-7). These results are in concordance with the prediction from the primary amino acid sequence that indicated a

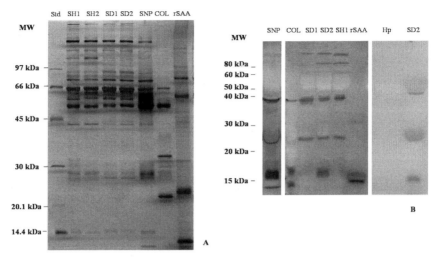

Figure 1. A: SDS-PAGE analysis of all studied samples, silverstained. B: Immunostaining of SAA proteins after SDS-PAGE analysis. Legend: Std: Amersham low molecular weight calibrators; SH1 and SH2: Serum of healthy pigs animals 1 and 2; SD1 and SD2: serum of diseased pigs animals 1 and 2; SNP: serum of newborn piglet; COL: sow colostrum; rSAA: recombinant Serum Amyloid A protein; Hp: purified pig haptoglobin. Molecular weights (MW) indicated in the left (Benchmark protein ladder, Invitrogen).

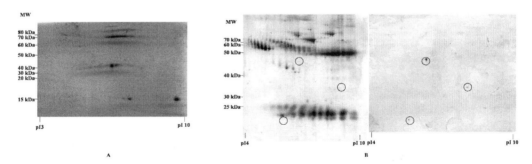

Figure 2. A: 2-DE analysis of rSAA, silverstained. B: 2-DE analysis of newborn piglet serum (SNP). The left image shows the SNP protein pattern by silverstaining and at the left image shows the SAA immunostaining of SNP. Position of positive spots are encircled. Molecular weights (MW) indicated on the left for each figure (Benchmark protein ladder, Invitrogen).

very alkaline protein as local SAAs[6], but also the presence of a less abundant neutral isoform. Furthermore, these results confirmed that SAA conformation is complex in the pig, and that oligomer formation affects SAA charge and mobility, as reflected from the more neutral pI of the rSAA oligomers. Same molecular weight SAA oligomers were also found in 2-DE SNP and

SD samples (Figs 2B and 3A, respectively). Neutral multimers (pI of 6.2-7) of approx. 45 to 47 kDa were found in serum from SD, as in SNP, where two additional spots were immunoreactive: 37 kDa (approx. pI of 8-9) and 23 kDa (approx. pI of 5.5). The obtained results indicated that pig circulating SAA isoforms were more alkaline than its counterparts from other species, but more acidic than calculated from its primary amino acid sequence, in concordance with previous findings[6]. This can be explained by the complexity of SAA composition in serum, in part due to its known complex processing and modification mechanisms[9]. Furthermore, these results confirm that SAA multimers and monomers can appear in pig serum. As formerly mentioned[4], the aggregation state of pig SAA might determine the functions of SAA in this species, which should be investigated in the future.

Two immunoreactive spots were identified in colostrum (Figure 3B), with molecular weights of approx. 13.5 and 15 kDa and a pI of more than 9. Main serum SAA pIs were around 7, while main colostrum SAA pIs were above 9 and not present in serum, suggesting local SAA production in the udder, and confirming previously published studies on SAA isoforms[6]. Authors believe that the 13 kDa immunoreactive protein found in colostrum both in SDS-PAGE and 2-DE should correspond with a fragment of SAA, but further studies must be carried out to confirm this.

There were some differences in SAA molecular weight estimation with SDS-PAGE and 2-DE, as well as there were also differences in pI estimation for positive spots compared with previously published one-dimensional IEF analysis[6]. However, sample treatments are substantially different before separation in each technique, which can affect the result of the

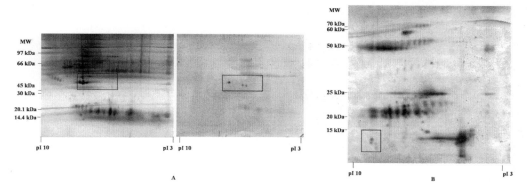

Figure 3. A: Representative 2-DE analysis of serum a diseased pigs (SD). The left image shows the SD protein pattern by silverstaining and at the left image shows the SAA immunostaining of SD. Position of positive spots are squared. Molecular weights (MW) indicated on the left for each figure (Amersham low molecular weight calibrators). B: 2-DE analysis of colostrum, silverstained. Position of positive spots are squared. Molecular weights (MW) indicated on the left for each figure (Benchmark protein ladder, Invitrogen).

analysis. Furthermore, alkaline proteins can have difficulties to enter the gel from rehydratation buffer, being cup-loading or paper strip-loading strategies more appropriate for optimal loading of basic proteins[10].

In conclusion, in the present study we confirmed that the pig circulating SAA monomer is highly alkaline, although pig SAA can be mainly found as more neutral multimers in serum from adult and newborn pigs. A very alkaline SAA monomer was found only in colostrum samples, indicating a local SAA production in the udder. Further studies must be conducted to determine the functional and physiopathological implications of the present findings.

References

1. Uhlar and Whitehead, 1999. Serum amyloid A, the major vertebrate acute-phase reactant. Eur J Biochem. 265(2):501-23.
2. Wang *et al.*, 2002. Murine apolipoprotein serum amyloid A in solution forms a hexamer containing a central channel. Proc Natl Acad Sci U S A. 99(25):15947-52.
3. Molenaar *et al.*, 2009. The acute-phase protein serum amyloid A3 is expressed in the bovine mammary gland and plays a role in host defence. Biomarkers. 14(1):26-37.
4. Raynes, 2006. Functions of acute phase proteins- with emphasis on serum amyloid A. Proceedings of the 6th European Colloquium on Acute Phase Proteins. Copenhaguen.
5. Niewold *et al.*, 2005. Chemical typing of porcine systemic amyloid as AA-amyloid. Amyloid. 12(3):164-6.
6. Soler *et al.*, 2011. Serum amyloid A3 (SAA3), not SAA1 appears to be the major acute phase SAA isoform in the pig. Vet Immunol Immunopathol. 141(1-2):109-15.
7. Fuentes *et al.*, 2011. Development of fast and simple methods for porcine haptoglobin and ceruloplasmin purification. Anales Veterinaria Murcia (In Press).
8. Laemmli, 1970. Most commonly used discontinuous buffer system for SDS electrophoresis. Nature. 222: 680-685.
9. Ducret *et al.*, 1996. Characterization of human serum amyloid A protein isoforms separated by two-dimensional electrophoresis by liquid chromatography/electrospray ionization tandem mass spectrometry. Electrophoresis. 17:866-76.
10. Görg *et al.*, 2009. 2-DE with IPGs. Electrophoresis. 1:122-32.

Comparative exoproteomics of *Staphylococcus epidermidis* of human and bovine origin to identify bacterial factors involved in adaptation into bovine host

Kirsi Savijoki[1], Pia Siljamäki[1,2], Niina Lietzén[2], Pekka Varmanen[1], Matti Kankainen[3] and Tuula A. Nyman[2]
[1]*Department of Food and Environmental Sciences, University of Helsinki, Finland*
[2]*Institute of Biotechnology, University of Helsinki, Finland; tuula.nyman@helsinki.fi*
[3]*Institute of Biomedicine, University of Helsinki, Finland*

Background

Gram-positive staphylococci include commensal and pathogenic bacteria that are adapted to different mammalian species, particularly to humans and cattle[1]. Among these species, *Staphylococcus epidermidis* is considered an opportunistic pathogen that ranks first among the causative agents of nosocomial infections and is the most common source of infection on indwelling medical devices[2]. *S. epidermidis* is also one of the major causative agents of bovine mastitis, which is a major economic burden in the dairy industry world-wide and the most common reason for antibiotic use in dairy cattle. It is also one of the most prevalent staphylococcal species found on human skin, which has led to speculations that bovine udder infections caused by *S. epidermidis* may be of human origin. The evolutionary success of this species depends on their remarkable ability to adapt to different environments and hosts[2], but the mechanisms by which these species select the host and establish successful infections is not well known.

Bacterial proteins expressed at cell surface (surfacome) or secreted into extracellular milieu (exoproteome) are expected to contribute to host-microbe interactions, and thus contain the most promising targets for diagnostic and vaccine innovations. To be infectious, *S. epidermidis* has to interact with the host and in this process both surfacome and exoproteome is believed to have a key role. We have previously exploited multigenome screen and analysis of several *S. epidermidis* genomes and complemented these analyses with total proteome comparison (intracellular proteome and surfacome) of a bovine mastitis causing *S. epidermidis* strain (mSP) and *S. epidermidis* ATCC12228 human strain with low infection potential [3] to search for host specificity determining factors in mSP. The goal of the present study was to compare exoproteomes of the mSP, the ATCC12228[4] and the infectious human *S. epidermidis* RP62A[5] strains to provide a deeper insight into the mechanisms by which the mastitis causing selects the host and establishes a productive infection.

Methods

Extraction and purification of the mSP, ATCC12228 and RP62A exoproteins was conducted according to the workflow presented in Figure 1. Briefly, supernatants were harvested from two independent cultures of the *S. epidermidis* strains mSP, ATCC12228 and RP62A by centrifugation and proteins were precipitated from filtered supernatants by TCA and purified using 2D CleanUp. To maximise the number of protein identifications we exploited high-throughput GeLC-MS/MS approach combining protein separation by SDS-PAGE and protein identification by LC-MS/MS (Figure 1). We have previously utilized this method for protein identification from grampositive bacteria[6], and in this study we followed the LC-MS/MS analysis.

Following LC-MS/MS all MS/MS spectra were searched against the ATCC12228 (2485 entries)[4], mSP (2530 entries)[3], and RP62A (2526 entries)[5] protein databases composed of

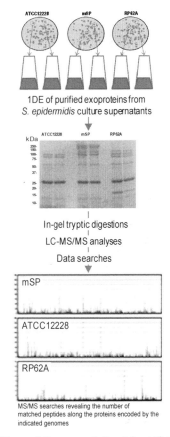

Figure 1. The GeLC-MS/MS workflow used for identifying Staphylococcus epidermidis *exoproteins.*

all chromosomal- and plasmid-derived ORFs using Mascot[7] and Paragon[8] search engines through the ProteinPilot™3.0 interface. The Compid tool was used to parse significant hits from the compiled Mascot and Paragon output files into tab delimited data files[9]. Protein identifications that had probability-based Mascot Mowse scores ≥50 (P<0.05) and/or Paragon Unused ProtScores ≥1.3 (P<0.05) were considered reliable high-confidence identifications. To estimate the false discovery rates (FDRs), all Mascot and Paragon searches were repeated using identical search parameters and validation criteria against the ATCC12228, mSP and RP62A decoy databases containing all protein sequences in both forward and reverse orientations. Sequences were reversed using the Perl decoy.pl script provided by Matrix Science (http://www.matrixscience.com/help/decoy_help.html. The FDR was calculated according to Elias *et al.*[10].

The theoretical isoelectric point (pI) and molecular weights (Mw) of the identified exoproteins were defined with ProMoST program (http://proteomics.mcw.edu/promost. html)[11]. Other information related to protein functional annotations, subcellular localization, presence of signal peptide, elements that indicate the mechanisms of cell-wall/membrane anchoring, and protein evolutionary counterparts (proteins that are homologs/paralogs and unique) were extracted elsewhere[3].

Results and future plans

The present study compares the exoproteins produced by three different *S. epidermids* strains of human (RP62A, ATCC12228) and bovine origin (mSP) to indicate mSP proteins with potential role in adaptation and interaction with the bovine host. We reliably identified 396, 487 and 484 proteins from the ATCC, RP62A and the mSP strains, respectively. The protein identification data were further assessed by virtual 2D proteome analysis (Figure 2), which

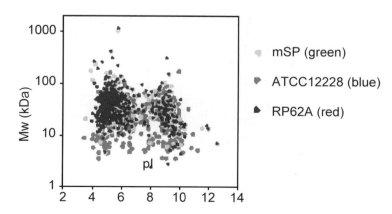

Figure 2. Virtual 2D proteome view of the identified mSP (green), ATCC12228 (blue) and RP62A (red) exoproteins.

shows an approximate bimodal distribution of the identified exoproteins and reveals that the protein extraction and identification is not biased.

The identified exoproteins felt into seven groups: (1) cell-wall associated proteins; (2) lipoproteins; (3) proteins that are secreted into the extracellular milieu; (4) proteins that are anchored to cell-wall via an unknown mechanism; (5) proteins carrying one or more trasmembrane domains (TMHMM); (6) proteins that are exported to culture media via yet an unknown mechanism; and (7) proteins that are with unknown function (Figure 3).

To define specifically secreted proteins, all identified proteins were next divided into exoproteins that were commonly and uniquely identified from each three strains (Figure 4). This analysis revealed that the mSP strain expressed 130 proteins that could not be identified from the commensal and infective human strains. The antigenicity of these uniquely secreted mSP

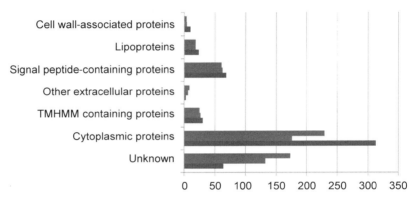

Figure 3. The number of the identified exoproteins divided according to their subcellular location and mechanisms of cell-wall or membrane anchoring.

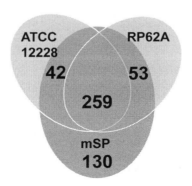

Figure 4. Venn diagram displaying the identified mSP exoproteins, which lack an identified counterpart from the ATCC12228 and RP62A human strains.

proteins will be further assessed by immunoblotting with antibodies from cows that have been experimentally infected with ATCC12228 and mSP[3].

This project is expected to enlighten the mechanisms employed by *S. epidermidis* for adaptation into different environments as well as for interaction with the bovine and/or human host. We also hope to discover new virulence factors as novel targets for diagnostics and vaccines.

References

1. Kloos WE, Schleifer KH. 1986. Genus IV. *Staphylococcus* Rosenbach 1884. In: Sneath, P.H.A., Mair, N.S., Sharpe, M.E., Holt, J.G. (Eds.), Bergey's manual of systematic bacteriology. Vol. 2, Williams & Wilkins, Baltimore, London, pp. 1013-35.
2. Otto M. 2009. *Staphylococcus epidermidis* – the 'accidental' pathogen. *Nature Rev* 7, 555-64.
3. Siljamäki P, Iivanainen A, Laine PK, Kankainen M, Simojoki H, Salomäki T, Pyörälä S, Paulin L, Auvinen P, Koskinen P, Holm L, Karonen T, Taponen S, Nyman TA, Sukura A, Kalkkinen N, Savijoki K, Varmanen P. Multigenome analysis and proteome profiling of *Staphylococcus epidermidis* strains to screen for host-specificity determining factors, *in preparation*.
4. Zhang YQ, Ren SX, Li HL, Wang YX, Fu G, Yang J, Qin ZQ, Miao YG, Wang WY, Chen RS, Shen Y, Chen Z, Yuan ZH, Zhao GP, Qu D, Danchin A, Wen YM. 2003. Genome-based analysis of virulence genes in a non-biofilm-forming*Staphylococcus epidermidis* strain (ATCC 12228). *Mol. Microbiol.* 49, 1577-1593.
5. Gill SR, Fouts DE, Archer GL, Mongodin EF, Deboy RT, Ravel J, Paulsen IT, Kolonay JF, Brinkac L, Beanan M, Dodson RJ, Daugherty SC, Madupu R, Angiuoli SV, Durkin AS, Haft DH, Vamathevan J, Khouri H, Utterback T, Lee C, Dimitrov G, Jiang L, Qin H, Weidman J, Tran K, Kang K, Hance IR, Nelson KE, Fraser CM. 2005. Insights on evolution of virulence and resistance from the complete genome analysis of an early methicillin-resistant *Staphylococcus aureus* strain and a biofilm-producing methicillin-resistant *Staphylococcus epidermidis* strain. *J Bacteriol* 187, 2426-2438.
6. Savijoki K, Lietzén N, Kankainen M, Alatossava T, Koskenniemi K, Varmanen P, Nyman TA. 2011. Comparative proteome cataloging of Lactobacillus rhamnosus strains GG and Lc705. J Proteome Res 10, 460-473.
7. Perkins DN, Pappin DJ, Creasy DM, Cottrell JS. 1999. Probability-based protein identification by searching sequence databases using mass spectrometry data. *Electrophoresis* 20, 3551-3567.
8. Shilov IV, Seymour SL, Patel AA, Loboda A, Tang WH, Keating SP, Hunter CL, Nuwaysir LM, Schaeffer DA. 2007. The Paragon Algorithm, a next generation search engine that uses sequence temperature values and feature probabilities to identify peptides from tandem mass spectra. *Mol Cell Proteomics* 6, 1638-1655.
9. Lietzén N, Natri L, Nevalainen OS, Salmi J, Nyman TA. 2010. Compid – a new software tool to integrate and compare MS/MS based protein identification results from Mascot and Paragon *J Proteome Res* 9, 6795-6800.
10. Elias JE, Gygi SP. 2007. Target-decoy search strategy for increased confidence in large-scale protein identifications by mass spectrometry. *Nat Methods 4, 207*-214.
11. Halligan BD, Ruotti V, Jin W, Laffoon S, Twigger SN, Dratz EA. 2004. ProMoST (Protein Modification Screening Tool): a web-based tool for mapping protein modifications on two-dimensional gels. *Nucleic Acids Res* 32:W638-44.

Methodology of proteomic mapping into milk infected by *Streptococus agalactiae*

William Mullen[1], Amaya Albalat[1], Aileen Stirling[1], Justyna Siwy[3] and Monika Johansson[2]
[1]Biomarkers and Systems Medicine, University of Glasgow, Glasgow, Scotland, United Kingdom;
William.Mullen@glasgow.ac.uk
[2]Department of Food Science, Swedish University of Agricultural Sciences, 75323 Uppsala,
Sweden
[3]Mosaique-Diagnostics, Hannover, Germany

Clinical proteomics ultimately aims at the improvement of a current clinical situation, generally via identification of proteins that show significant changes associated with disease. Early diagnosis of disease is a key factor for the successful outcome of the treatment. The technology platforms for proteome analysis have progressed considerably over the last few years. Driven by these advancements in technology, the number of studies on the analysis of the proteome/ peptidome, with the aim of defining clinically relevant biomarkers, has substantially risen (Figure 1).

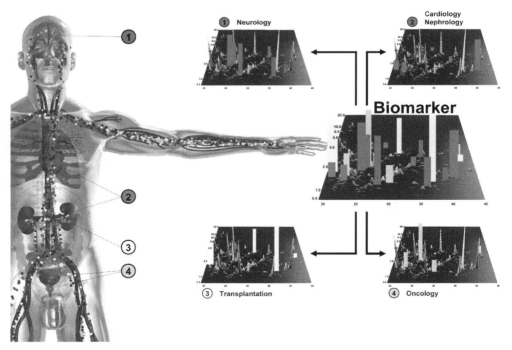

Figure 1. Diseases currently with validated urinary biomarker fingerprints.

We have developed methodology for the analysis of urinary clinical biomarkers of a number of chronic diseases including coronary heart disease (CAD) and chronic kidney disease (CKD) using capillary electrophoresis mass spectrometry to detect pre-symptomatic signs of these diseases and to monitor treatment progression (Figure 2).

It is of interest to evaluate different mastitis pathogens and their strains also from the milk quality perspective to gain knowledge about their ability to degrade the economical important proteins in the milk.

The analytically methodology compares the peptide profiles produced by case and control samples using advanced statistical analysis of the quantitative peptide data. As this is comparative proteomic the sample collection, treatment and extraction must be kept as simple as possible so as not to introduce any possible variables to the peptide content of the sample. This is one of the reasons we do not use blood as the biomarker source. A large number of variables can be generated during the collection and storage of the blood sample prior to centrifugation. In addition, due to the wide dynamic range of protein concentrations in plasma and the highly complex composition of it, there is always the need for a separation or enrichment method to allow the proteins/peptides of interest to be measured. These factors introduce a number of potential errors that make biomarker research in blood an almost impossible task.

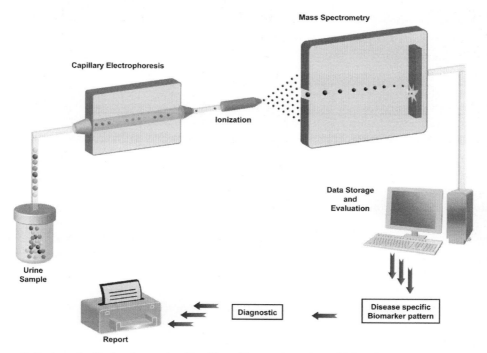

Figure 2. Proteomics Technology using Capillary Electrophoresis and Mass Spectrometry CE-MS.

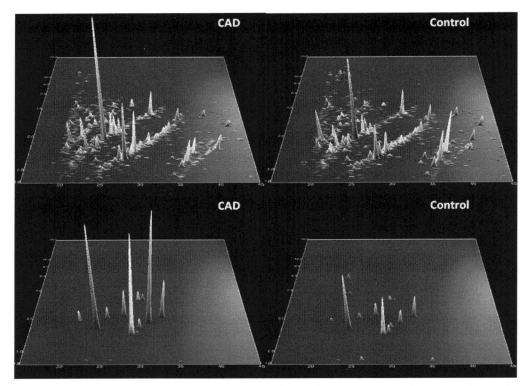

Figure 3. Peptide fingerprints of CAD and control samples and the peptides that make up the biomarker model.

In the investigation carried out as part of the COST action STSM we wanted to know if the urinary proteomics techniques could be used to investigate UHT milk that had been incubated with known strains of *Streptococus agalactiae* and for known durations of incubation. As the methodlgy measures intact peptides it is a simple matter to track this back to the original protein. UHT milk was chosen as a sample media as it allowed for a more controlled sample sets and increased number of known samples replicated without any possible interference from unknown bacterial infections..

In associated, non proteomic based studies, it was shown that there were broad variations of proteolytic activity caused by the different strains of the same bacteria. The question we were asking was whether we could adapt the urinary methodology for early detection of infection of milk samples by a rapidly spreading mastitis pathogen *S. agalactiae* and, if successful, could it be use to distinguish between different strains belonging to the same phylogenic group of the bacteria and also to follow the time course of the infection.

Once a biomarker model has been generated samples will be analysed by LC-MS/MS in an attempt identify the peptides that form the model revealing an insight into the proteins and proteases involved.

The results of the analysis of these samples will be presented.

Acknowledgements

Figures 1, 2 and 3 courtesy of Mosaique-Diagnostics GmBH.

The work was funded by a COST action STMS to Monika Johansson and William Mullen.

Part III
Proteomics in animal production

Proteome analysis of muscles longissimus dorsi of Hungarian Merino and Tsigai sheep breeds

Levente Czeglédi, Krisztina Pohóczky, Gabriella Gulyás, Beáta Soltész and András Jávor
University of Debrecen, Centre for Agricultural and Applied Economic Sciences, Institute of
Animal Science. 138. Boszormenyi Street, 4032 Debrecen, Hungary; czegledi@agr.unideb.hu

More than 80% of sheep flocks in Hungary belong to the breed Hungarian Merino (Nagy *et al.*, 2011). The first animals were imported to the country in 1770s. This breed is bred for dual purpose, its meat and wool are the products for the market, but meat production is more significant and profitable. In the breeding programmmes the focus is on increasing prolificacy and lamb rearing ability, while in case of wool production the better quality is the aim.

The Tsigai breed originated from Asia, with the first animals arrived in Hungary at the end of 18th century from Romania and the Balkan countries. It has spread widely and become popular, because it is not only dual purpose but its milk production is significant beside meat and wool (Komjáthy *et al.*, 1996). This breed had a larger ratio in the Hungarian sheep population in the last century, but only small flocks still exist at present time, most of them in the region between Danube and Tisza rivers.

The aim of this study was to investigate the differences in the muscle proteome of Hungarian Merino and Tsigai sheep breeds by gel based proteomic tools.

Animals and methods

10 rams were involved in the study, four Hungarian Merino and six Tsigai. Animals were born in January and slaughtered at the weight of 40 kg and age of 4.5 month in average. Housing and feeding conditions were the same for each animal. They were fed ad libitum with concentrate and hay, rearing in a stable, kept in small groups.

Muscle samples were harvested into cryotubes, with 2 g per sample in three replicates, 20 min. after slaughter and placed immediately into liquid nitrogen, and then were kept at -80 °C until subsequent analysis.

Protein samples were prepared from sheep muscle tissue as follows: tissue samples were placed in liquid nitrogen and ground thoroughly to a very fine powder with a mortar and pestle. The tissue powder (50 mg) was transferred to sterile tubes containing 700 µl of lysis buffer (8 M urea, 2 M thiourea, 2% (w/v) CHAPS, 50mM DTT, 0.2% (v/v) Bio-Lyte 4/6 and 6/8 ampholytes at a ratio 1:2) and 0.08% (v/v) protease inhibitor cocktail (Fermentas). The mixture was incubated for 60 min. on ice with occasional vortexing and centrifuged at 15,000xg for 40 min., the supernatant was collected and stored at -80 °C until further analysis. The protein content was determined using a protein assay kit (Bio-Rad, Hercules, CA, USA) with bovine serum albumin (BSA) as standard.

For the first dimension (isoelectric focusing) of two-dimensional gel electrophoresis, 17 cm immobilized pH gradient (IPG) strips (pH 5-8, linear, Bio-Rad) were rehydrated in 300 µl of rehydration buffer (2 M thiourea, 8 M urea, 2% (w/v) CHAPS, 50 mM DTT, 0.2% (v/v) Bio-Lyte 4/6 and 6/8 ampholytes at a ratio 1:2, 0.002% (w/v) Bromphenol Blue) for 15 h at room temperature. 100 µg of protein was loaded for the analytical gels, whereas 700 µg was loaded for the preparative gels. Isoelectric focusing was conducted in Protean IEF Cell (Bio-Rad). Low voltage (250 V) was applied for 20 min. The voltage was gradually increased to 10,000 V over 2.5h, and was maintained at that level until a total of 50,000 Vh. The current limit was adjusted to 50 mA per strip, and the run was carried out at 20 °C. Focused IPG strips were equilibrated for 10 min in 6 M urea, 20% (v/v) glycerol, 2% (w/v) SDS, 50 mM Tris pH 8.8 and 2% (w/v) DTT, and then for an additional 10 min. in the same buffer except that DTT was replaced by 2.5% (w/v) iodoacetamide. After equilibration, proteins were separated in the second dimension using Protean II XL vertical electrophoresis system (Bio-Rad). Second dimension was performed on 160x200 mm, 13% polyacrylamide gels. Gels were run at 16 mA at the first 30 min. and then 24 mA until the bromphenol blue dye marker reached the end of the gels. A cooling system provided constant 20 °C running temperature. The gels were stained with silver staining as described by Shevchenko *et al.* (1996). Stained gels were matched and analyzed with Delta2D software (Decodon ™).

Results

315 protein spots were identified on gels of which 27 showed different expression. Twenty of 27 were overexpressed in Hungarian Merino muscle samples and seven had higher spot volume% in Tsigai breed (Figure 1). Eight spots were at $P<0.01$, seventeen at $P<0.05$, two at $P=0.05$ significant. In a similar study for proteome analysis, longissimus lumborum of two different pig breeds were analyzed by polyacrylamide gel electrophoresis and the work resulted in 23 differentially expressed proteins (D'Alessandro *et al.*, 2011). Zhao and co-workers (2010) found 36 proteins comparing the proteome of cattle breeds.

The identification of differentially expressed proteins is in progress by matrix-assisted laser desorption/ionization – time of flight mass spectrometry.

Acknowledgements

The work was supported by the TÁMOP-4.2.2/B-10/1-2010-0024 and TÁMOP 4.2.1-08/1-2008-003 projects. The project is co-financed by the European Union and the European Social Fund.

pH 5   pH 8

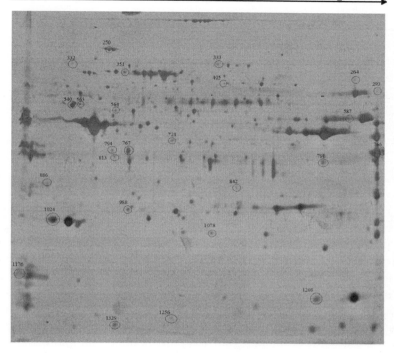

Figure 1. 2D PAGE of musculus longissimus dorsi. 'O' indicates proteins show different expression depending on breed.

References

D'Alessandro A., Marrocco C., Zolla V., D'Andrea M., Zolla L. (2011) Meat quality of the longissimus lumborum muscle of Casertana and Large White pigs: Metabolomics and proteomics intertwined. Journal of Proteomics 75. (2). 610-627.

Komjáthy Gy., Maknics Z., Márkó J., Mentes K., Racskó P., Tímár L. (1996) /Knowledge in Animal Breeding, Book for Agricultural Education/ Gazdaképzés tankönyve, Állattenyésztési ismeretek, Mezőgazdasági Szaktudás Kiadó Budapest 287-290.

Nagy Zs., Németh Á., Mihályfi S., Toldi Gy., Gergátz E., Holló I. (2011) The short history of Hungarian sheep breeding and Hungarian Merino breed. Acta Agraria Kaposvariensis 15. (1). 19-26.

Shevchenko A., Wilm M., Vorm O., Mann M. (1996) Mass spectrometric sequencing of proteins silver-stained polyacrylamide gels. Analytical Chemistry 68. (5). 850-858.

Zhao M., Basu U., Dodson M., Basarb J., Guan L. (2010) Proteome differences associated with fat accumulation in bovine subcutaneous adipose tissues. Proteome Science 8. 14.

Muscle proteomics profiles in sheep: the effect of breed and nutritional status

André M. de Almeida[1,2], Rui M.G. Palhinhas[1], Tanya Kilminster[3], Tim Scanlon[3], Johan Greeff[3], Chris Oldham[3], John Milton[4], Ana V. Coelho[1] and Luís Alfaro Cardoso[2]
[1]*Instituto de Tecnologia Química e Biológica, Oeiras, Portugal; aalmeida@fmv.utl.pt*
[2]*Instituto de Investigação Científica Tropical and CIISA – Centro Interdisciplinar de Investigação em Sanidade Animal, Lisboa, Portugal*
[3]*Department of Agriculture and Food, Western Australia, Perth, WA, Australia*
[4]*University of Western Australia, Perth, WA, Australia*

Introduction

Seasonal Weight Loss (SWL) is the most relevant constraint to animal production in the tropics and Mediterranean countries (Almeida *et al.*, 2006). In fact, during the dry season characteristic of such climates animals may lose up to 20% of their initial weight, significantly affecting their productive performances with strong economic implications (Almeida *et al.*, 2006).

The study of the physiological and biochemical mechanisms by which domestic animal breeds respond to SWL is of capital interest with important implications in animal selection (Almeida, *et al.*, 2010). The sheep (*Ovis aries*) is used worldwide for research in production and physiology. The study of the proteome has brought important information on several physiological mechanisms including those underlying SWL, although little information is available for production animals. The objective of this study is to determine differential protein expression at the level of the Gastrocnemius muscle in three different sheep breeds with different levels of tolerance to SWL: the Australian Merino (AM) an European Breed, considered to have low levels of adaptation to nutritional stress; the Damara (DA) a Southern Africa fat tail sheep breed with high levels of adaptation to nutritional stress and finally the Dorper (DO), a composite breed (Dorset horn X Persian Black Head) of South African origin and considered to have intermediate levels of tolerance to SWL (Almeida, 2011).

Material and methods

Male lambs (6 months old) of three different sheep breeds, Australian Merino, Dorper and Damara, showing different levels of tolerance to weight loss (low, medium and high, respectively) were used (n=4). Per breed, 2 experimental groups were established (control and weight loss). The trial lasted a total of 42 days. At the end of the trial, animals were euthanized in a commercial abattoir and gastrocnemius sampled and frozen at -80°C until further analysis.

The extraction protocol has been previously described (Almeida *et al.*, 2010), using an Ultraturrax anda 8 M urea, 2 M thiourea and 4% (w/v) extraction buffer. Extracts were quantified with the 2D Quant kit (GE Heathcare, Uppsala, Sweden). Two-dimensional gel electrophoresis was conducted using 24 cm pH 3-10 immobiline dry strips and colloidal coomassie blue staining. Gels were analyzed using the Same Spots software (Non-Linear Dynamics, Newcastle, UK). Spots of interest for identification were in-gel digested with trypsin and proteins identified using MALDI-TOF/TOF (Puerto *et al.*, 2011).

Results and discussion

After 42 days of the experimental period, lambs subjected to food restriction, lost approximately 10-15% of the initial body weight, whereas animals in the control group had an increase of approximately 10%. After sacrifice, excision of the gastrocnemius muscle and two-dimensional electrophoresis, gels such as the one depicted in Figure 1 were obtained. Upon gel analysis, and as marked in Figure 1, a total of 22 spots were selected for identification, of which 16 were successfully identified.

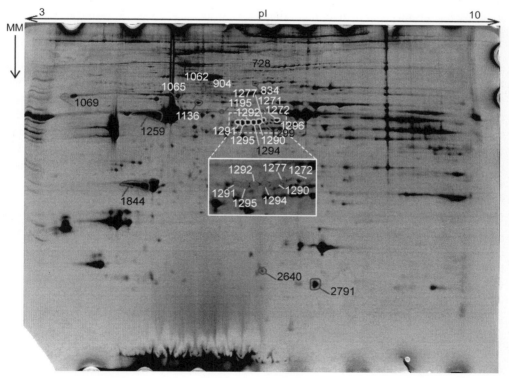

Figure 1. Reference Gel showing 22 spots with differential expression and selected for identification. MM – Molecular Mass; pI – Isoelectric point.

The following proteins were found to be differentially expressed between at least two experimental groups: histidine triad nucleotide-binding protein, myosin light chain, desmin, actin, cytochrome b-c1 complex subunit 1 mitochondrial, troponin T, and phosphoglucomutase-1. According to a previously defined classification (Almeida *et al.*, 2009), identified proteins were classified as structural (16%), metabolic (23%) and of the contractile apparatus (32%).

Contrary to preliminary experiments with Wild and New Zealand White (NZW) rabbits (Almeida *et al.*, 2010), the number of spots showing differential expression is much lower in the present experiment. Several reasons may justify this fact; one of the most important could be the severity of food restriction that was much higher in the case of the rabbit experiment (20% of the initial weigh loss). Interestingly, in the rabbit experiment, the bulk of protein expression differences were more a consequence of breed than of the nutritional factor itself. In the present experiment, only one spot (#1259) identified as actin showed a different profile between the controls of Merino and Dorper animals, higher in the first. Such results were somehow unexpected as, being selected for higher meat production levels, Dorper sheep tend to have a higher muscular development than Merino sheep, in a situation contrary to the one reported for wild and NZW rabbits (Almeida *et al.*, 2010).

Another spot showing an interesting pattern of expression was spot #2640, identified as histidine triad nucleotide-binding protein. For this spot, a higher level of expression was detected between Merino restricted and Damara restricted animals, whereas both Dorper groups and Damara control and Merino control had a similar expression level. The fact that the Merino is considered to show higher susceptibility to nutritional stress than the Damara that was selected in a much harsher environment it could be inferred that expression levels of Histidine Triad nucleotide-binding protein could be considered as a putative candidate to the lack of tolerance to seasonal weight loss.

Results presented in this study seem to point out to a similar protein expression profiles as a consequence of breed and nutritional status. Nevertheless, and in order to assess such breed differences, it would be interesting to conduct a follow up gel analysis focusing only in the three control groups. In this study, we have identified essentially majorly expressed proteins, with special relevance to structural and contractile apparatus proteins. It is likely however that the definition of markers of tolerance to SWL would not be reflected in these two classes of proteins. It would therefore be interesting to conduct this study using methods that prove to be of higher capacity to detect minor proteins, namely DIGE or fluorescent dyes.

Acknowledgements

Authors would like to thank project POCI/CVT/57820/2004 from FCT – *Fundação para a Ciência e a Tecnologia* (Lisboa, Portugal) and the Department of Agriculture and Food of the Government of the Western Australia (Perth, Australia) for financing this Research. AM Almeida acknowledges financial support from the *Ciência 2007* program of the FCT.

References

Almeida, A.M. (2011). The Damara in the context of Southern Africa fat tailed sheep breeds. Tropical Animal Health and Production 43: 1427-1441.

Almeida, A.M., A. Campos, R. Francisco, S. van Harten, L.A. Cardoso and A.V. Coelho (2010). Proteomic investigation of the effects of weight loss in the gastrocnemius muscle of wild and New Zealand white rabbits via 2D-electrophoresis and MALDI-TOF MS. Animal Genetics 41: 260-272.

Almeida, A.M., A. Campos, S. van Harten, L.A. Cardoso and A.V. Coelho (2009). Establishment of a proteomic reference map for the gastrocnemius muscle in the rabbit (*Oryctolagus cuniculus*). Research in Veterinary Science 87: 196-199.

Almeida, A.M., L.M. Schwalbach, H.O. de Waal, J.P. Greyling and L.A. Cardoso (2006). The effect of supplementation on productive performance of *Boer* goat bucks fed winter veld hay. Tropical Animal Health and Production 38: 443-449.

Almeida, A.M., L.M.J. Schwalbach, L.A. Cardoso and J.P.C. Greyling (2007). Scrotal, testicular and semen characteristics of young Boer bucks fed winter veld hay: the effect of nutritional supplementation. Small Ruminant Research 73: 216-220

Puerto, M., A. Campos, A. Prieto, A. Cameán, A.M. Almeida, A.V. Coelho, V. Vasconcelos (2011). Differential protein expression in two bivalve species, *Mytilus galloprovincialis* and *Corbicula fluminea*, exposed to *Cylindrospermopsis raciborskii* cells. Aquatic Toxicology 101: 109-116.

Cellular and molecular-large scale features of fetal perirenal and intermuscular adipose tissues in bovine

M. Bonnet, H. Taga, B. Picard and Y. Chilliard
INRA, UR1213 Herbivores, 63122 Saint-Genès-Champanelle, France;
Muriel.Bonnet@clermont.inra.fr

In ruminants, adequate fetal and post-natal development of adipose tissues (AT) is a major challenge to promote metabolic adaptation both at birth for neonate survival (especially in sheep), and in adult life for productive efficiency during the pregnancy-lactation cycle of dairy females or for carcass yield and quality of meat animals. The factors that determine AT mass in adult mammals are not fully understood. Both epidemiological and fetal programming studies in ruminants suggest that fetal growth of ATs impacts their postnatal growth and therefore the body fat mass or adiposity (Bonnet *et al.*, 2010). A prerequisite to understand the mechanisms of fetal programming of AT growth is to improve our knowledge on the mechanisms underlying the AT growth in fetus, which are still very poorly known whatever the species. In this context, our objective was to identify the cellular and molecular mechanisms that sustain the fetal AT growth in bovine and to study its variability.

Methods

We combined measurements of chemical composition, cellularity, histology, enzyme activities, gene expression and proteomics to describe the ontogeny of perirenal and intermuscular AT in bovine at 110, 180, 210 and 260 days post conception (dpc) in Blond d'Aquitaine (n=3 per age) and Charolais (n=5 per age) breeds as well at 260 dpc only in Holstein breeds (Taga *et al.*, 2011, 2012 a, b). These breeds were chosen for their differences in adiposity in the post-natal life.

For the proteomics analysis, proteins extracted from the perirenal AT of Blond d'Aquitaine and Charolais fetuses at 110, 180, 210 and 260 dpc, were separated in the first-dimensional electrophoresis using 4-7 pH range. These acid proteins were separated in the second dimension on separating 12% polyacrylamide gels and were stained with G250 colloïdal coomassie blue. Protein spot detection and volume quantification were realized with SameSpots V.3 software. We detected 261 well-resolved protein spots in the gel image that were common to the four fetal ages (Figure 1).

Peptide mass fingerprintings of trypsin-digested proteins were determined with a voyager-DE pro Maldi-Tof mass spectrometer and compared to *Bos taurus* database of NCBInr using MASCOT 2.2 software, which allowed the identification of 235 spots corresponding to 143 proteins. Data from the proteomics experiments were analyzed using EDGE, an open-source software package designed to perform significance analysis of time-course expression data

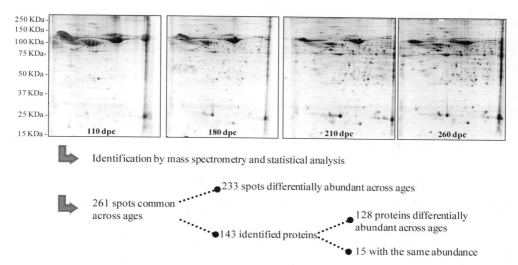

Figure 1. 2-DE of perirenal AT proteins at 110, 180, 210 and 260 days post conception (dpc) in fetuses. Gels were loaded with 700 µg of total protein extracted from Charolais or Blond d'Aquitaine fetuses.

(Storey *et al.*, 2005). Proteins whose abundance changed over time (whatever the breed) were considered significant at *P*<0.05.

Results

Between 110 and 260 dpc (38 and 90% of gestation length, respectively), the increase in the weight of perirenal AT resulted from an increase in the adipocytes volume and mainly number (Figure 2). The increases in adipocyte volume and mainly number were accompanied by changes in the abundance of 128 proteins among the 143 proteins identified and common to the four fetal ages studied (Figure 1). Among the identified proteins, some of them have never been described in the AT and may contribute to hyperplasia of adipose precursors, by controlling cell cycle progression, apoptosis and/or by delaying adipocyte differentiation. The age of 180 dpc seems to be a pivotal age for the transition between proliferation and differentiation of adipocyte progenitors. An increase in the abundance of many proteins involved in differentiation and in the increase in adipocyte volume was observed from 180 dpc (Taga *et al.*, 2012a).

As soon as 180 dpc up-regulations were unexpectedly observed in the β-subunit of ATP synthase, which is normally bypassed in brown AT, as well as in aldehyde dehydrogenases ALDH2 and ALDH9A1, which were reported to be predominantly expressed in mouse white AT relatively to brown AT (Forner *et al.*, 2009). From 210 dpc, we observed an increase in the mRNA abundances of UCP1 and DIO2, which are hallmarks of brown adipocytes. These

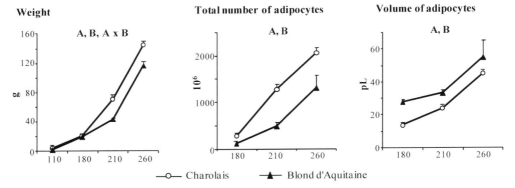

Weight   **Total number of adipocytes**   **Volume of adipocytes**

—o— Charolais   —▲— Blond d'Aquitaine

Figure 2. Weight of perirenal AT (g), total number of perirenal adipocytes and adipocyte volume (pL) at 110 (only for AT weight), 180, 210 and 260 days post conception in Charolais and Blond d'Aquitaine fetuses. Results are means ± SEM. B, A: significant effect of breed or age (P<0.001); B x A: significant interaction (P<0.05).

results, associated with the presence of numerous unilocular adipocytes and few multilocular adipocytes from 180 dpc (Figure 3), shows that fetal ATs up to 260 dpc have molecular and cellular characteristics of white AT in addition to those of brown AT (Taga *et al.*, 2012a). Our results challenge the commonly accepted concept that fetal perirenal AT in bovine is brown, and highlight that perirenal AT is a heterogeneous tissue up to 260 dpc in bovine fetuses.

At 260 dpc, perirenal and intermuscular AT from Holstein fetuses, compared with Charolais and Blond d'Aquitaine, had larger adipocytes, and higher expression of genes involved in lipogenesis, lipolysis and endocrine function. The total number of adipocytes in the perirenal

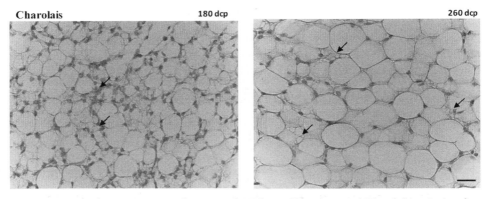

Figure 3. Hematoxylin-eosin-stained perirenal AT from Charolais and Blond d'Aquitaine fetuses at 180 and 260 days post conception (dpc). Arrows indicate multilocular brown adipocytes among the unilocular white adipocytes. Bar = 25 μm.

AT was however lower in Holstein, so that the weight of perirenal AT did not differ between the three breeds (Taga *et al.*, 2011). Lastly, whatever the breed, perirenal and intermuscular AT had similar metabolic and cellular characteristics (Taga *et al.*, 2011, 2012b).

Conclusion

The cellular and molecular features observed in our study challenge current concepts on the largely brown nature of bovine foetal AT (based on histological and metabolic features previously reported a few days before or after birth for perirenal AT), and strongly suggest that fetal bovine perirenal and intermuscular AT have much more in common with white than with brown AT. Our data highlights that two of the major AT sites contain mainly unilocular putatively white adipocytes two to three weeks before birth, and thus emphasizes the potential limitation in heat production for the survival of pre-term ruminants born in harsh environments. This raises questions over the specific location of brown adipocytes and/or the mechanisms by which perirenal and intermuscular AT could acquire the metabolic properties required for thermogenesis before birth. A better knowledge of the balance between brown and white fat cells could allow to better control not only thermogenesis and survival in perinatal period, but also the energy efficiency during the productive life of the ruminants. Elsewhere, proteins regulating the number of fat cells remain to be studied in other models (postnatal bovine, other species), as possible predictors of the proliferation of fat cells and growth potential of AT.

References

Bonnet M., Cassar-Malek I., Chilliard Y., Picard B. 2010. Ontogenesis of muscle and adipose tissues and their interactions in ruminants and other species. Animal 4: 1093-1109.

Forner F, Kumar C, Luber CA, Fromme T, Klingenspor M, Mann M. 2009. Proteome Differences between Brown and White Fat Mitochondria Reveal Specialized Metabolic Functions. Cell. Metab. 10:324-335.

Storey JD, Xiao W, Leek JT, Tompkins RG, Davis RW. 2005. Significance analysis of time course microarray experiments. Proc. Natl. Acad. Sci. USA 102:12837-2842.

Taga H., Bonnet M., Picard B., Zingaretti M. C., Cassar-Malek I., S. Cinti, and Y. Chilliard., 2011. Adipocyte metabolism and cellularity are related to differences in adipose tissue maturity between Holstein and Charolais or Blond d'Aquitaine fetuses. J. Anim. Sci. 89: 711-721.

Taga H., Chilliard Y., Picard B., Zingaretti M. C., Bonnet M., 2012b. Foetal bovine intermuscular adipose tissue exhibits histological and metabolic features of brown and white adipocytes during the last third of pregnancy. Animal doi:10.1017/S1751731111001716 (In Press).

Taga H., Y. Chilliard, B. Meunier, C. Chambon, B. Picard, C. Zingaretti, S. Cinti, and M. Bonnet., 2012a. Cellular and molecular-large scale features of fetal adipose tissue: is bovine perirenal adipose tissue brown? J. Cell. Physiol. 227:1688-1700.

Vegetable based fish feed changes protein expression in muscle of rainbow trout (*Oncorhynchus mykiss*)

Flemming Jessen, Tune Wulff, Jeanett Bach Mikkelsen, Grethe Hyldig and Henrik Hauch Nielsen
National Food Institute, Technical University of Denmark, 2800 Kgs. Lyngby, Denmark;
fjes@food.dtu.dk

Introduction

Feed production for aquaculture of carnivore fish species relies heavily on protein and lipid from the limited resources of wild fish and other sea living organisms. Thus the development of alternative feeds replacing fish meal and oil with alternatives of e.g. vegetable origin is important for a sustainable production of fish from aquaculture (Glencross *et al.* 2007). Such a change in feed ingredients will affect the metabolic pathways in fish and during the last years genomic and proteomic techniques have contributed in obtaining a better understanding of the involved mechanisms (Panserat & Kaushik 2010; Martin *et al.* 2003; Vilhelmsson *et al.* 2004). A consequence of fish meal and fish oil replacement in the feed may also be a change of eating quality (Johnsen *et al.* 2011; Lesiow *et al.* 2009).

Material and methods

Rainbow trout were reared to approx. 450 g in two groups given different diets for 12 weeks: a traditionally control diet (C) based on marine oil and protein and a diet (V) based exclusively on vegetable products. Both feed contained 42% protein and 26% fat.

Muscle samples (0.5 g) for a gel based proteome analysis were withdrawn just after slaughter. Water soluble proteins were subjected to 2-dimensional gel electrophoresis (2-DE) (Wulff *et al.* 2008), gels were Coomassie stained (Rabilloud & Charmont 2000) and Progenesis SameSpots (Nonlinear Dynamics) was used for image analysis. The differentially expressed spots were identified using tandem mass spectrometry.

The textural attributes flaky, firm, juicy, fibrousnesses, and oiliness of the fish were measured after chilled storage in ice for 7 days by sensory profiling of heat treated filets as described by Baron *et al.* (Baron *et al.* 2009).

Multivariate data analysis, partial least square (PLS) regression, was performed using the software 'The Unscrambler[®]' (v. 9.1, Camo, Norway). A jack-knife method (Martens & Martens 2000) for selection of spots with significant ($P<0.05$) regression coefficients was included.

Results and discussion

2-DE comparison of the fish muscle revealed 39 spots (Figure 1) that were significantly (Student t-test; $P<0.05$) different between the two feeding groups C (n=8) and V (n=7). The major part (25 spots) of the 39 spots represented up-regulated proteins whereas 14 spots represented down-regulated proteins in fish from the V group. The provisionally identified spots (14) are indicated by numbers in the gel (Figure 1) and given by name, function, and direction of expression change in fish from the V group in Table 1. It was found that intake of the vegetable based diet, among others, influenced the expression of muscle proteins involved in lipid binding/transport, protein turnover, and binding of different ions.

The sensory profiling showed a significant (two-tailed t-test; $P<0.05$) difference between the two groups in the textural attribute firm, with the fish fed the vegetable based diet being the firmest. However, there was a big within group variation in the sensory data and this variation might be the reason for not finding other significant textural differences between the two groups.

To investigate if the overall (independent of the feeding groups) textural variation was reflected in the 2-DE gel, data multivariate data analysis (PLS regression) was performed correlating spot volumes with the textural attributes. This resulted in the finding of 46 spots that individually or in combination with other spots correlated to one or more of the textural attributes, except to oiliness. In total eleven of these spots correlating to firm (7 spots), flaky (4 spots), and juicy (2 spots) were also among the 39 spots significantly differing between the feeding groups C and V indicating that the vegetable based feed might as well have an influence on other textural attributes than firm.

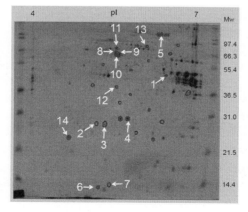

Figure 1. Muscle protein expression differences in rainbow trout fed the two type of feed. Proteins (39) of interest based on Student t-test (P<0.05) are marked. The 2-DE gel is a representative gel of water soluble proteins from rainbow trout muscle. Mw is given i kDa.

Table 1. Identified differential expressed proteins in fish feed the two type of feed.

No.	Protein name	Effect[1] of vegetable based feed	Function
1	6-phosphogluconate dehydrogenase	↑	Pentose shunt
2-32	**Apolipoprotein A-I-1 precursor**	↓	**Lipid/cholesterol transport**
4	Carbonic anhydrase 1	↑	One-carbon metabolic process
5	**eEF1A2 binding protein-like**	↑	**Protein synthesis (translation)**
6	Fatty acid-binding protein	↑	Lipid binding and transport
7	Fatty acid-binding protein	↓	Lipid binding and transport
8-9	**Hemopexin-like protein**	↑	**Metal binding**
10	T-complex protein 1 subunit theta	↑	Protein folding
11	Thimet oligopeptidase	↓	Proteolysis
12	Selenoprotein J	↑	Selenium binding
13	Transferrin	↓	Iron binding
14	Translationally-controlled tumor protein	↑	Calcium binding; microtubule stabilization

[1] Arrow pointing up shows up-regulated protein expression in fish fed the vegetable based diet, and *vice versa*.
[2] Proteins in **bold** were among those also correlating to the textural attributes.

It is concluded that changing from the marine based to the vegetable based diet resulted in a firmer texture of rainbow trout muscle and influenced the expression of several muscle proteins, among those proteins that were also correlated to the textural properties flaky and juicy.

References

Baron CP, Hyldig G & Jacobsen C 2009 Does Feed Composition Affect Oxidation of Rainbow Trout (Oncorhynchus mykiss) during Frozen Storage? Journal of Agricultural and Food Chemistry 57 4185-4194.

Glencross BD, Booth M & Allan GL 2007 A feed is only as good as its ingredients – a review of ingredient evaluation strategies for aquaculture feeds. Aquaculture Nutrition 13 17-34.

Johnsen CA, Hagen O, Adler M, Jonsson E, Kling P, Bickerdike R, Solberg C, Bjornsson BT & Bendiksen EA 2011 Effects of feed, feeding regime and growth rate on flesh quality, connective plasma hormones in farmed Atlantic salmon (Salmo salar L.). Aquaculture 318 343-354.

Lesiow T, Ockerman HW & Dabrowski K 2009 Composition, Properties and Sensory Quality of Rainbow Trout Affected by Feed Formulations. Journal of the World Aquaculture Society 40 678-686.

Martens H & Martens M 2000 Modified Jack-knife estimation of parameter uncertainty in bilinear modelling by partial least squares regression (PLSR). Food Quality and Preference 11 5-16.

Martin SAM, Vilhelmsson O, Medale F, Watt P, Kaushik S & Houlihan DF 2003 Proteomic sensitivity to dietary manipulations in rainbow trout. Biochimica et Biophysica Acta-Proteins and Proteomics 1651 17-29.

Panserat S & Kaushik SJ 2010 Regulation of gene expression by nutritional factors in fish. Aquaculture Research 41 751-762.

Rabilloud T & Charmont S 2000 Detection of proteins on two-dimensional electrophoresis gels. In Proteome research: Two-dimensional gel electrophoresis and identification methods., pp 107-126. Ed Thierry Rabilloud. Springer Verlag, Berlin Heidelberg.

Vilhelmsson OT, Martin SAM, Medale F, Kaushik SJ & Houlihan DF 2004 Dietary plant-protein substitution affects hepatic metabolism in rainbow trout (Oncorhynchus mykiss). British Journal of Nutrition 92 71-80.

Wulff T, Jessen F, Roepstorff P & Hoffmann EK 2008 Long term anoxia in rainbow trout investigated by 2-DE and MS/MS. Proteomics 8 1009-1018.

The proteomic insight of the Italian dry cured ham manufacturing

Gianluca Paredi[1], Samanta Raboni[1], Anna Pinna[2], Giovanna Saccani[2], Roberta Virgili[2] and Andrea Mozzarelli[1,3]

[1]Department of Biochemistry and Molecular Biology, Interdepartmental Center Siteia.Parma, University of Parma, Parma, Italy; gianluca.paredi@nemo.unipr.it
[2]Stazione Sperimentale per l'Industria delle Conserve Alimentari, Parma, Italy
[3]National Institute of Biostructures and Biosystems, Rome, Italy

Introduction

Proteomics is a powerful tool for the identification of proteins that are characteristic of the transformation of meat in industrial processes and might be diagnostic of the quality of the final product. Parma dry cured ham is a typical Italian food product, manufactured according to established technological steps mainly based on subjective empirical observations and traditional recipes. Meat salt intake, dehydration and water activity decrease are some of the key events that lead to a food product that can be stored at room temperature for several months. Specifically, the development of the sensory and quality traits of the cured ham is accomplished by a process that includes salting, resting, drying and maturation:

1. During salting phase sections of pork posterior legs are covered with a weighted amount of dry and wet salt and stored in cold rooms (1-3 °C) at high relative humidity (RH>80%) for two-three weeks. This step promotes salt solubilisation/penetration inside meat mass and affects product safety, sensory, quality and nutritional properties. A protein rich exudate is produced due to the salt-induced breakage of muscle cells.

2. During the resting phase, air circulation at low temperature (2-4 °C) and decreased relative humidity enhance meat dehydration.

3. During the drying and maturation steps (12 months and over), in which hams are stored under controlled conditions of temperature (14-18 °C), air circulation and %RH, dehydration, protein solubilisation, proteolysis, lipolysis and generation of volatile molecules take place allowing typical sensory characteristics to be achieved.

The aim of this study was the proteomic analysis of the salting exudate dripping from meat over the salting period in order to understand the effect of salt on meat protein, the salt diffusion rate and exudate protein composition.

Materials and methods

30 fresh pork meat pieces of domestic heavy pig, average weight 13.18±0.15 kg, were selected in a local slaughterhouse, according to the pH range 5.7-5.9. Hams were divided in two classes: 20 hams were pressed (12 kg/ham for 24 h) perpendicular to the fat-free muscle mask.

Pressure was applied to achieve a decrease both in ham thickness and in shape variability. The remaining 10 hams were left untouched preserving the standard shape. Meat was salted in two steps, using a mixture of wet and dry salt (NaCl). Hams were manually salted using two kinds of salt (a mixture of 2- and 3-mm grain size): wet salt (nearly 15% added water) was rubbed on ham rind, while dry salt was used for the unskinned ham part. Five per cent salt (calculated on ham weight) was used for the first salting, while an addition of 2.8% salt was made in the second salt. Salted hams were placed in a cold room operating at 80-90% RH, at 1-3 °C. At the end of the first salting phase (sixth day) hams were desalted and submitted to a second salting, performed in the same environmental conditions of the first salting for 12 days. The whole salting treatment lasted 18 days. To collect the exudate dripped from hams during salting, six hams (4 pressed and 2 standard-shape) were placed into six large vessels. The exudate was collected from each ham after 1, 5 and 18 days. Exudate samples were stored at -80 °C before processing. First, the exudates were centrifuged in order to remove insoluble components. The protein concentration was then assayed with the Bradford method. Due to the high salt concentration in the samples a protein isolation step was carried out on 200 µl exudate aliquots by quantitative acetone precipitation.

Results and discussion

Protein concentration was found to remain constant between day 1 and 5 at a value of about 3 mg/ml, and significantly to decrease at day 18 (Figure 1).

A comparison of the proteomes at day 1, 5 and 18 was carried out with 1-DE (Figure 2). The profile of the exudates showed a higher amount in two protein bands with low molecular weights at day 1 than at day 5 and 18. Interestingly, three bands showed a higher concentration at day 18 compared to day 1 and 5. A 2-DE analysis of the three proteomes was carried out in the pH range 3-10. Results will be presented and discussed.

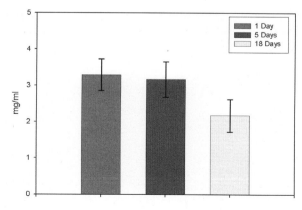

Figure 1. Average protein concentration in exudates collected at day 1, 5 and 18.

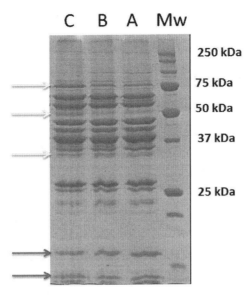

Figure 2. SDS-PAGE of exudates collected at 1 (A), 5 (B) and 18 (C) days. The black arrows mark protein bands more intense in the 1 day sample. The light grey arrows mark protein bands more intense in the 18 days sample.

Conclusions

Our preliminary proteomic analysis of exudates from the salting phase in the preparation of dry cured ham indicates a differential release of specific proteins as a function of salting time. These proteins will be identified by mass spectrometry.

The presence of high salt concentration in dry cured hams is associated with health concerns, thus calling for alternative or modified manufacturing strategies. Our proteomic investigations will help in re-directing dry cured ham manufacturing to obtain high quality products at lower salt concentration.

IGF-1 free circulating protein and its gene-expression in linseed-rich diet quail

A. Karus[1], H. Tikk[2], A. Lember[2], V. Karus[1] and M. Roasto[3]
[1]Department of Chemistry, Institute of Veterinary Medicine and Animal Sciences, Eesti Maaülikool, Tartu, Estonia; avo.karus@emu.ee
[2]Department of Small Farm Animal and Poultry Husbandry, Institute of Veterinary Medicine and Animal Sciences, Eesti Maaülikool, Tartu, Estonia
[3]Department of Food Hygiene and Control, Institute of Veterinary Medicine and Animal Sciences, Eesti Maaülikool, Tartu, Estonia

Introduction

Quail are highly important poultry in regional agricultural poultry meat and egg production (Tikk *et al.*, 2009, 2011). Linseed oil and linseed cake provide components to increase n3-polyunsaturated fatty acid (n3 PUFA) content in poultry products. As described in olive oil and conjugated linoleic acid example, the modification of feed fat composition may result in adverse effects in bird physiological status (Aydin *et al.* 2001; Karus *et al.*, 2007). Insulin-like growth factors have been well studied and the effect of IGF-1 has been shown to play a significant role in chicken muscle development in all phases (Kocamis and Killefer 2003; Saprōkina *et al.*, 2009). The aim of the study was to examine IGF-1 gene-expression and IGF-1 protein content in different feeding conditions in quail blood.

Material and methods

Animals and experimental design

Trials were carried out according to Estonian Animal Protection Act (13.12.2000/RT I 2004). Six hundred 21-day-old Estonian quail (*Coturnix coturnix*) chicks were randomly assigned to three treatment groups. The birds were fed a starter diet up to the 21st days of age followed by a finishing diet from day 21 to day 42. The birds were fed either basal diet or the basal containing linseeds or linseed cake. Content of finishing diets is shown in Table 1.

In the experiment (from day 21 up to the age of 42 days) quails were fed diets containing 4% rapeseed oil and 8% linseed (Flaxseed) or 10% linseed cake (in the first and second trial group, respectively). The control group consumed ordinary feed containing 4% rapeseed oil. At the age of 42 days, blood samples were collected from 10 chickens (5 male and 5 female birds of the body weight similar to the sex's average) in each group. Blood was collected from *V. jugularis* into disposable non-heparinized test tubes for IGF-1 testing and into EDTA-diNa tubes for mRNA studies. Blood serum was separated by centrifugation at 2,000x*g* for 10 min and was then frozen (-24 °C) for analysis. Whole blood samples were stabilized using RNA/

Table 1. Comparison of n3 polyunsaturated fatty acid (n3 PUFA) content in quail diets.

Group / Feed	Feed fat %	n3 PUFA in 1 g feed (mg)	n3 PUFA in daily ration (mg)
Group A Linseed (8% linseed)	9.72	13	390
Group B Linseed cake (10% linseed cake)	5.94	14	420
Control group	4.73	5	150
Linseed	32.77	157	
Linseed cake	13.46	65	

DNA Stabilization Reagent for Blood/Bone Marrow (Roche Applied Science), and then frozen (-24 °C) for analysis.

Methods of laboratory analyses

IGF-1 in blood serum was measured by RIA (DSL) kit. Roche mRNA Isolation Kit for Blood/Bone Marrow was used for mRNA purification from stabilized blood samples. The LightCycler RNA Master SYBR Green I kit (Roche Applied Science) was used for hot start one-step RT-PCR in glass capillaries using the LightCycler Instruments and SYBR Green I dye as the detection format as described earlier (Karus *et al.*, 2007). GAPDH gene was used for IGF-1 mRNA relative quantification as housekeeping-gene. Data was analyzed using the statistical programs SYSTAT 11.0.

Results and discussion

The variation of free circulating insulin-like growth factor concentration in quail blood serum and IGF-1 expression in leukocytes was relatively high. Both IGF-1 concentration in blood serum and its mRNA relative concentration in leukocytes a tendency to decrease in groups with linseed modified diets. One has to mention, that IGF-1 level in plasma increases approximately six- to seven-fold from hatching to day 21 and then remains relatively stable with tendency to increase, but hepatic IGF-1 mRNA level increases in the same period only three-fold (Giachetto *et al.* 2004) and there was no clear correlation between the growth of the birds and hepatic IGF-1 mRNA expression and plasma IGF-1 levels. Distribution of birds in our experiments according to dietary manipulations is more spectacular than average IGF-1 content given in Karus *et al.* (2007) (Figure 1).

In comparison of two parameters: IGF-1 protein content and IGF-1 relative gene-expression, there is clear advantage to measure native free protein since the distribution of birds by

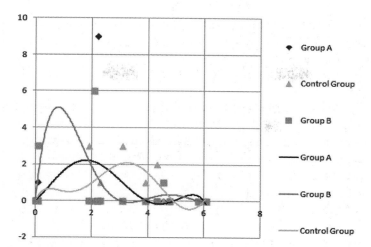

Figure 1. IGF-1 protein content in quail blood.

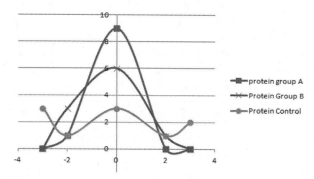

Figure 2. Birds distribution by IGF protein content. Average=0 (± SD).

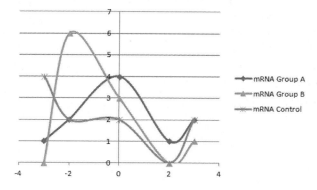

Figure 3. Birds distribution by IGF mRNA   relative content Average=0 (± SD).

mRNA level is significantly asymmetric, while native protein content is still close to normal distribution (Figure 2 and 3).

Conclusion

Both IGF-1 concentration in blood serum and its mRNA relative concentration in leukocytes showed the tendency to decrease in groups with linseed modified diets. Since, the free IGF-1 has less asymmetric distribution profile in all bird groups and basic metabolic effect of regulatory proteins depends on proteins themselves, the measurement of protein content may have significant advantages before traditional gene-expression studies.

References

Aydin R., M.W. Pariza, and M.E. Cook, 2001: Olive Oil Prevents the Adverse Effects of Dietary Conjugated Linoleic *Acid on Chick Hatchability and Egg Quality. Journal of Nutrition.* 131, 800-806.

Giachetto P.F., E.C. Riedel, E. Gabriel, M.I.T. Ferro, S.M.Z. Di Mauro, M. Macari, and J.A. Ferro, 2004: Hepatic mRNA expression and plasma levels of insulin-like growth factor-I (IGF-I) in broiler chickens selected for different growth rates. Genetics and Molecular Biology, 27, 1, 39-44.

Karus A., Saprõkina Z., Tikk H., Järv P., Soidla R., Lember A., Kuusik S., Karus V., Kaldmäe H., Roasto M., Rei M., 2007: Effect of Dietary Linseed on Insulin-Like Growth Factor-1 and Tissue Fat Composition in Quails. Archiv für Geflügelkunde / European Poultry Science, 71(2), 81-87.

Kocamis H., and J. Killefer, 2003: Expression Profiles of IGF-I, IGF-II, bFGF and TGF-β2 Growth Factors during Chicken Embryonic Development. Turk. J. Vet. Anim. Sci. 27, 367-372.

Saprõkina Z., Karus A., Kuusik S., Tikk H., Järv P., Soidla R., Lember A., Kaldmäe H., Karus V., Roasto M., 2009: Effect of dietary linseed supplements on ω-3 PUFA content and on IGF-1 expression in broiler tissues. Agricultural and Food Science, 18(1), 35-44.

Tikk H., Lember A., Karus A., Tikk V., Piirsalu M., 2009: Eestis kasvatatud vutibroilerite lihajõudlus ja liha keemiline koostis = Meat performance and meat chemical composition of quail broilers in Estonia. Agraarteadus: journal of agricultural science: Akadeemilise Põllumajanduse Seltsi väljaanne, XX(2), 47-59.

Tikk H, Lember A., Tikk V., Piirsalu M., 2011: Egg performance dynamics of Estonian Quail in 1987-2010. J. of Agricultural Science / Agraarteadus, XXII(2), 71-78.

A longitudinal proteomic approach to investigate liver metabolism in periparturient dairy cows with different body fat mobilization

Christine Schäff[1], Dirk Albrecht[2], Harald M. Hammon[1], Monika Röntgen[1], Cornelia C. Metges[1] and Björn Kuhla[1]
[1]*Research Unit Nutritional Physiology 'Oskar Kellner', Leibniz Institute for Farm Animal Biology (FBN), Wilhelm-Stahl-Allee 2, 18196 Dummerstorf, Germany; b.kuhla@fbn-dummerstorf.de*
[2]*Institute of Microbiology, Ernst-Moritz-Arndt-University, F.-L.-Jahn-Strasse 15, 17487 Greifswald, Germany*

In late gestation dairy cows start to mobilize body reserves primarily glycogen, protein and fat depots. The mobilization of fat reserves results in an increased plasma concentration of free fatty acids and in a fat deposition in the liver accompanied by changes of hepatic lipid metabolism after calving. According to the hepatic oxidation theory (HOT) [1], increased hepatic fatty acid oxidation is suggested to depress feed intake during early lactation. Therefore, we investigated the expression of enzymes participating in hepatic beta-oxidation in dairy cows with different body fat mobilization to identify correlations with dry matter intake (DMI). Daily *ad libitum* DMI was recorded for 19 multiparous German Holstein cows (2nd to 4th lactation, >10,000 kg/305d in at least one previous lactation) housed in a tie stall from week 7 before until 5 weeks after calving. Cows were fed three different total mixed rations according to their physiological state, in far-off dry period (week -7 to -4, 5.87 MJ NEL and 128 g nXP/kg DM), close-up dry period (week -3 till calving, 6.49 MJ NEL and 137 g nXP/kg DM) and lactation (week 1 to 5, 7.06 MJ NEL and 163 g nXP/kg DM). The liver of each cow was biopsied at five time points [days -34, -17, 3, 18, 30 rtp (relative to parturition)] and at slaughter (day 40 rtp) and the obtained samples were immediately frozen in liquid nitrogen. Total liver fat content (LFC) was determined by the CN-method [2] and used as a parameter of body fat mobilization in early lactation. According to the average *post partum* (days 3, 18, and 30 rtp) LFC, cows were grouped into high-mobilizing (HMC; LFC >24.4% DM; n=10) and low-mobilizing (LMC; LFC <24.0% DM; n=9) cows. Also, samples (50 mg) of frozen liver tissue were crushed to a fine powder and extracted for a proteomic approach as described earlier for feed restriction-induced fatty liver [3]. Briefly, extracts were applied to 2-dimensional gel electrophoresis followed by colloidal Coomassie staining (Figure 1). For spot detection and quantification using the normalized spot volume, 2D gels were processed using Delta 2D software (DECODON). Data were analysed by the Mixed Model of SAS with LFC and time relative to parturition as fixed effects. All spots that were differentially expressed between groups ($P<0.05$; Mixed Model of SAS, LFC and time as fixed effects) were picked, tryptic digested and analysed on a 5800 MALDI TOF/TOF analyser.

The subsequent NCBInr and Swiss-Prot database search yielded 170 identified protein spots. Among them we obtained 10 spots representing isoforms of beta-oxidative enzymes, namely

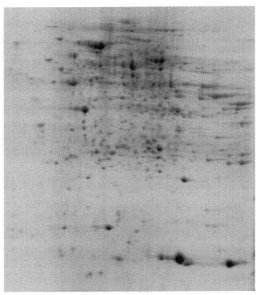

Figure 1. Representative Colloidal Coomassie stained 2-D gel of an individual liver biopsy obtained from a cow 34 days before calving. Proteins were horizontally separated on an IPG gel strip (pH 3-10) and vertically on a 12.5% SDS PAGE gel (20x20x0.1 cm).

long chain acyl-CoA dehydrogenase, medium chain acyl-CoA dehydrogenase, short chain acyl-CoA dehydrogenase, 3-hydroxy acyl-CoA dehydrogenase and 3-ketoacyl-CoA thiolase. Interestingly, acyl-CoA dehydrogenases and 3-hydroxy acyl-CoA dehydrogenase were higher expressed in HMC than in LMC. By contrast, all spots representing 3-ketoacyl-CoA thiolase were lower expressed in HMC in comparison to LMC. These results strongly suggest that the first and the third enzyme of the beta-oxidative machinery were up-regulated in HMC cows because of the higher fat mobilisation as compared to LMC. The reduced expression of 3-ketoacyl-CoA thiolase (the fourth enzyme of the beta-oxidative pathway) in HMC, however, indicates a limited availability of CoA in these animals. Thus, it appears that up-regulation of enzymes located upstream of thiolase is not sufficient to prevent fatty liver and that complete fatty acid oxidation is necessary to avoid fatty liver. These findings further suggest that HMC oxidize more fatty acids – albeit not completely to AcCoA – and produce more ketone bodies than LMC. Also, increased but incomplete fatty acid oxidation is clearly associated with the lower DMI/kg body weight observed in HMC as compared to LMC. This is in agreement with HOT but if the higher fatty acid oxidation causes the lower DMI/kg BW or whether the lower DMI/kg BW effects higher fatty acid oxidation after calving remains to be determined.

Acknowlegements

The study was supported by Deutsche Forschungsgemeinschaft (DFG; KU 1956/4-1)

Farm animal proteomics

References

1. Allen MS, Bradford BJ, Oba M. Board Invited Review: The hepatic oxidation theory of the control of feed intake and its application to ruminants. J Anim Sci. 2009; 87: 3317-34.
2. Kuhla S, Klein M, Renne U, Jentsch W, Rudolph PE, Souffrant WB. Carbon and nitrogen content based estimation of the fat content of animal carcasses in various species. Arch Anim Nutr 2004; 58: 37-46.
3. Kuhla B, Albrecht D, Kuhla S, Metges CC. Proteome analysis of fatty liver in feed-deprived dairy cows reveals interaction of fuel sensing, calcium, fatty acid, and glycogen metabolism. Physiol Genomics. 2009; 37: 88-98.

The effect of colostrum intake on blood plasma proteomic profiles of newborn lambs

Lorenzo Enrique Hernández-Castellano[1], André Martinho de Almeida[2,3], Miguel Ventosa[2], Antonio José Morales-delaNuez[1], Ana Varela Coelho[2], Diego Martell Jaizme[1], Noemí Castro Navarro[1] and Anastasio Argüello Henríquez[1]
[1]*Department of Animal Science, Universidad de Las Palmas de Gran Canaria, Arucas, Gran Canaria, Spain; lhernandezc@becarios.ulpgc.es*
[2]*Mass Spectrometry Laboratory, Instituto de Tecnología Química e Biologica, Universidad Nova de Lisboa, Oeiras, Portugal*
[3]*Centro de Veterinária e Zootencia, Instituto de Investigação Científica Tropical & CIISA – Centro Interdisciplinar de Investigação em Sanidade Animal, Lisboa, Portugal*

Introduction

The consumption of colostrum by newborn ruminants plays a fundamental role in the acquisition of passive immunity. In fact, due to the characteristics of ruminant's syndesmochorial placenta, the transfer of immunoglobulins from the dam to the fetus is very limited and not able to ensure the survival of the newborn (Castro *et al.*, 2009).

Yvon *et al.* (1993) have described the importance of colostral proteins absorption in order to achieve a correct passive immunity transfer; however it has not been described which proteins are absorbed by lambs from colostrum and their role in the passive immunity transfer.

The objective of this study is to determine which proteins increased/decreased expression profiles after colostrum intake in the newborn lamb during the first 12 hours after birth. We also aim to describe which of them are related to either the passive immunity transfer or lamb immune system development, using an approach based on 2-D electrophoresis and MALDI-TOF/TOF.

Material and methods

Two groups, of six lambs each, of the Canaria breed were fed with sheep colostrum at different times, according to the experimental design described in Table 1. This part of the experiment was conducted at the Veterinary Faculty of the Universidad de Las Palmas de Gran Canaria (Canary Islands, Spain). One group received three colostrum meals (colostrum group), at 2, 14 and 26 hours after birth. The other group (no colostrum group) was fed with colostrum at 14 and 26 hours after birth. At the end of the colostral period (26 hours after birth) each animal (from both groups) took the same amount of immunoglobulins (Ig×s) per live body weight (4mg of IgG/Kg) from colostrum. Blood samples were collected at 2 and 14 hours after birth and the obtained plasma was frozen at -80 °C until further analysis.

Table 1. Experimental design.

Group name	Number of animals	Hypothetical proteins in blood due to the colostrum intake.	
		2 hours after birth	14 hours after birth
Colostrum	6	No	Yes
No colostrum	6	No	No

The second part of the experiment was carried out at the Instituto de Tecnología Química e Biologica (Oeiras, Portugal). Plasma samples were treated with the ProteoMiner® Protein Enrichment Kit (Bio-Rad), desalted with 2D-Clean-up® kit (GE Healthcare) and quantified with 2D-Quant® kit (GE Healthcare). Samples were then subjected to DIGE two-dimensional gel electrophoresis. 2DE was conducted using 24 cm pH 3-10 immobiline dry strips (GE Healthcare). Gel analysis was performed using Progenesis SameSpots software (Nonlinear). In order to excise the selected spots, gels were stained with colloidal Coomassie and spots were manually picked, subjected to trypsin digestion and proteins identified using MALDI-TOF/TOF as previously described (Lamy *et al.*, 2009).

Results and discussion

After gel analysis, a total of 11 spots resulted with differential expression, showing 10 of them an increase in the colostrum group and 1 of them a decrease in the no colostrum group (Figure 1). These spots were selected for identification using MALDI-TOF/TOF. We have successfully identified a total of 6 which were increased in the colostrum group.

Identified proteins include: Apolipoprotein A-IV (spots 565 and 572), Plasminogen (spot 201), Serum Amyloid A (spot 726) and Fibrinogen (spots 490 and 747). It is very important to understand how colostrum modified the plasma proteome in newborn lambs, so the function of differential expressed proteins will be described individually.

It has been suggested that Apolipoprotein A-IV (Apo A-IV) plays an important role at early stage of life, modulating the enterocyte lipid transport efficiency in fatty foods, namely colostrum (Bisgaier *et al.*, 1985) and it also has an immunomodulatory effect against external agents (Vowinkel *et al.*, 2004).

Another identified protein was Plasminogen which plays an important role in the dissolution of Fibrin blood clots in order to prevent thrombosis. However, the structural homology of Apo A-IV and Plasminogen endows Apo A-IV with the capacity to bind to Fibrin and to

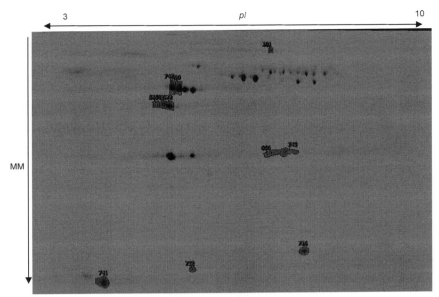

Figure 1. Reference Gel showing 11 spots with differential expression and selected for identification. MM – Molecular Mass; pI – Isoelectric point.

membrane proteins of endothelial cells and Monocytes, and thereby to increase the thrombosis rate (Angles-Cano, 1997). Probably this increased amount of Plasminogen could thwart the unhealthy effects produced by high amounts of Apo A-IV.

Another apolipoprotein identified in this study was the Serum Amyloid A which takes part of the acute phase of the inflammation, being increased to 1000-fold its level under normal conditions (Schultz and Arnold, 1990). This protein is a potent chemoattractant for human Leukocytes, such as Monocytes, Neutrophils and T Lymphocytes (Su *et al.*, 1999), so an increase of this protein could be beneficial for the newborn lamb at this stage.

The last analyzed protein was identified as Fibrinogen gamma chain. This is one of the three pair of chains that compose Fibrinogen, which in turn is the precursor of Fibrin, the most abundant component of blood clots, being also increased during the acute phase of the inflammation. Moreover, Yamada *et al.* (2002) studied differences in low abundance proteins between cow colostrum and milk, finding that some of them were only present in the colostrum, namely Fibrinogen. This could probably explain the fold increase of this spot in the colostrum group.

As a conclusion, further studies will be needed to identify the rest of the differentially expressed spots. However, with this study it has been proved that colostrum is not only important because it contains immunoglobulins such as IgG, IgM or IgA, that are necessary for a correct passive immunity transfer, but also because it contains several proteins which play an important role in the protection of the newborn lamb at this early stage of life, namely Apolipoprotein A-IV (apo A-IV), Plasminogen, Serum Amyloid A and Fibrinogen.

Acknowledgements

Authors would like to thank grant AGL2009-11944 from MICINN -Ministerio de Ciencia e Innovación (Spain) and AM Almeida the Ciência 2007 program of Fundação para a Ciência e a Tecnologia (Lisboa, Portugal).

References

Angles-Cano E (1997). Structural basis for the pathophysiology of lipoprotein(a) in the athero-thrombotic process. Brazilian Journal of Medical and Biological Research 30, 1281-1290.

Bisgaier CL, Sachdev OP, Megna L, Glickman RM (1985). Distribution of apolipoprotein A-IV in human plasma. Journal of Lipid Research 26, 11-12.

Castro N, Capote J, Morales-delaNuez A, Rodríguez C, Argüello A (2009). Effects of newborn characteristics and length of colostrum feeding period on passive immune transfer in goat kids. Journal of Dairy Science, 92, 1616-1619.

Hook TE, Odde KG, Aguilar AA, Olson JD (1989). Protein effects on fetal growth, colostrum and calf immunoglobulins and lactation in dairy heifers. Journal of Animal Science 67, 539.

Lamy E, da Costa G, Santos R, Capela e Silva F, Potes J, Pereira A, Coelho AV, Sales Baptista E (2009). Sheep and goat saliva proteome analysis: A useful tool for ingestive behavior research?. Physiology & Behavior 98, 393-401.

Schultz DR, Arnold PI (1990). Properties of four acute phase proteins: C-reactive protein, serum amyloid A protein, alpha 1-acid glycoprotein and fibrinogen. Seminars in Arthritis and Rheumatism 20,129-147.

Su SB, Gong W, Gao JL, Shen W, Murphy PM, Oppenheim JJ, Wang JM (1999). A seven-transmembrane, G protein-coupled receptor, FPRL1, mediates the chemotactic activity of serum amyloid A for human phagocytic cells. Journal of Experimental Medicine 189,395-402.

Vowinkel T, Mori M, Krieglstein CF, Russell J, Saijo F, Bharwani S, Turnage RH, Davidson WS, Tso P, Granger DN, Kalogeris TJ (2004). Apolipoprotein A-IV inhibits experimental colitis. Journal of Clinical Investigation 114, 260- 269.

Yamada M, Murakami K, Wallingford JC, Yuki Y (2002). Identification of low abundance proteins of bovine colostral and mature milk using two-dimensional electrophoresis followed by microsequencing and mass spectrometry. Electrophoresis 23, 1153-1160.

Yvon M, Levieux D, Valluy M, Pélissier J, Mirand PP (1993). Colostrum protein digestion in newborn lambs. Journal of Nutrition 123,586-592.

Oxidative stress and acute phase response associated to changes in housing from pen to individual stalls in reproductive sows

Anna Marco-Ramell[1], Raquel Peña[1], Laura Arroyo[1], Raquel Pato[1], Yolanda Saco[1], Lorenzo Fraile[2,3] and Anna Bassols[1]

[1]*Departament de Bioquímica i Biologia Molecular, Universitat Autònoma de Barcelona, Spain; anna.bassols@uab.cat*
[2]*Centre de Recerca en Sanitat Animal (CReSA), UAB-IRTA, Campus de la Universitat Autònoma de Barcelona, 08193 Bellaterra, Barcelona, Spain*
[3] *Universitat de Lleida, 25198 Lleida, Spain*

Background

Under field conditions, gilts are usually housed in pens of 10-15 sows by group, when they arrive into the production facility from the multiplication unit. After spending there the quarantine period, they are moved into the mating room, where they are housed in small and individual stalls where the mating process takes place. This individual housing system could be stressful for the animals [1]. The aims of our study were twofold. The first one is to confirm that this change of housing from pen to stalls is really stressful for gilts and the second one is to find new welfare/chronic stress markers if it were possible in this experimental setting.

Experimental design

A diagram of the experimental design in this trial is presented in Figure 1. The study began in the quarantine facility (day 1 of the trial) where all the included animals were sampled (60 sows) and they were moved to individual stalls this same day. Afterwards, two groups of animals were established. The first one (study group) included 15 sows that were sampled at day 3, 4 and 5 of the trial. The third group included 45 sows which were samples in groups of 15 sows at day 3, 4 and 5 of the trial. This experimental design will allow us to distinguish the effect due to housing from the stress possible induced by the sampling process Thus, this control group could be defined as a handling marker.

Saliva (Salivette™) and blood (Vacutainer™) samples were collected in the mouth or through jugular vein sampling, respectively at the days described previously. Finally, it was obtained, plasma serum, erythrocytes lysate and leucocytes from blood samples. Saliva was used without any additional laboratorial management.

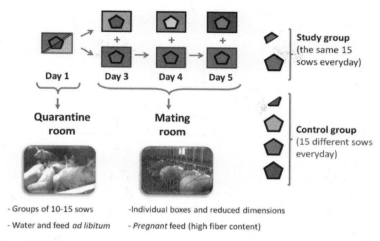

Figure 1. Experimental design.

Methods

Health, nutritional and oxidative stress markers, cortisol and acute phase proteins were determined using enzymatic or colorimetric assays on an Olympus automatic analyzer, or through commercial ELISAs. The proteomic approach was performed with 2-DE DIGE (fluorescent labeling), followed by mass-spectrometry (MALDI-TOF and LC-MS/MS) for protein identification.

Results and discussion

In general terms, gilts were apparently healthy by daily visual analysis and this clinical observation was confirmed through biochemical markers. However, some of them were altered in the study group, showing that daily blood collection caused muscular damage.

Nutritional markers measurements showed that animals have decreased the food intake on day 3, but normal food intake was recovered at the end of the study. The salivary cortisol increased in both groups throughout the experiment.

Oxidative stress markers varied due to the housing change that suffered the gilts. Superoxide dismutase (SOD) and glutathione peroxidase (GPx) were measured in erythrocyte lysate, and total glutathione (GSH) in whole blood. SOD, GPx and GSH increased the following day after the housing change (from 1 to 3) in both groups (study and control group), but they slightly decrease in the study group from day 3 to 5 whereas this decrease was not observed in the control group (Figure 2).

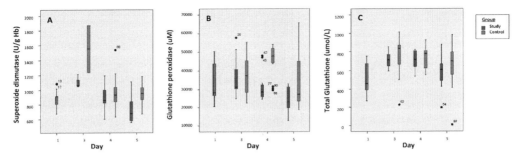

Figure 2. Oxidative stress markers' levels throughout the experiment: A) Superoxide dismutase; B) Glutathione peroxidase; C) total glutathione.

In the proteomic approach, Serum proteome of animals belonging to study and control groups on day 1, day 3 and day 5 were analyzed to look for proteins that could vary their concentrations throughout the experiment with the same tendency on both groups. The goal was to find a stress biomarker associated to change in housing and, not associated to the blood sampling procedure.

27 differential spots were selected for protein identification, according to an ANOVA *P*<0.05 and a ±1.2 fold variation observed between days. The preliminary MS identifications are represented in Table 1.

The preliminary MS results revealed that two positive acute phase proteins, haptoglobin and the inter-alpha-trypsin inhibitor heavy chain H4 (ITIH4 or Pig-MAP), modified their levels when the animals were housed in individual boxes in comparison with the blood sampling procedure. Antithrombin, which is usually measured in coagulation analysis, and apolipoprotein A-IV, which is involved in the cholesterol metabolism, also increased associated to the change in housing of the gilts.

Table 1. Preliminary mass-spectrometry identifications.

Spot #	Identification	ANOVA	Fold	acces #
669	Serpin peptidase inhibitor (antithrombin) (*Sus scrofa*)	7.012E-03	1.2	gi|194018664
802	Haptoglobin precursor (*Sus scrofa*)	1.679E-05	1.2	gi|47522826
1188	Inter-alpha-trypsin inhibitor heavy chain H4 precursor (*Sus scrofa*)	1.104E-02	1.4	gi|48374067
1253	Apolipoprotein A-IV precursor (*Sus scrofa*)	1.156E-05	1.3	gi|47523830

Haptoglobin and Pig-MAP were validated with a colorimetric assay and a commercial ELISA, respectively. Both parameters increased on day 3 of the trail, but they decreased on the following days in both groups (study and control). It was also measured another acute phase protein, the protein C-reactive (CRP), which increased markedly in the study group, but not in the control one. Thus, this result suggests that this increase of CRP at day 4 and 5 is probably due to blood sampling and not to stress due to change in housing (Figure 3).

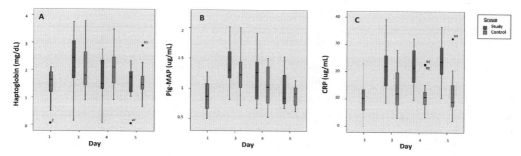

Figure 3. Acute phase protein levels along the experiment: (A) Haptoglobin; (B) Pig-MAP; (C) CRP.

Conclusions

The sows subjected to a change in their housing, from a quarantine pen to an individual box, showed a biochemical profile similar to a pro-inflammatory status, with increased levels of oxidative stress markers and acute phase proteins. Moreover, the inclusion of the control group, totally necessary in any experimental setting and critical in stress studies, allowed observing that this pro-inflammatory status is longer in animals suffering daily handling for blood sampling.

References

1. Barnett, JL. *et al.* Applied Animal Behaviour Science, 32 (1991) 23-33.

Two-dimensional proteomics as a tool to evaluate nutritional effects in farmed fish

Nadège Richard[1], Paulo Gavaia[1], Mahaut de Vareilles[1,2], Tomé S. Silva[1,3], Odete Cordeiro[1], Pedro M. Rodrigues[1] and Luis E.C. Conceição[1]

[1]*CCMAR, Centro de Ciências do Mar do Algarve, universidade do Algarve, Campus de Gambelas, 8005-139 Faro, Portugal; pmrodrig@ualg.pt*
[2]*Department of Biology, University of bergen, Norway*
[3]*DTU Food, Fødevareinstituttet, Copenhagen, Denmark*

The use of proteomics in fish research is at a relatively early stage compared to terrestrial vertebrates. It has already been applied in the field of fish nutrition, enabling to point out metabolic changes occurring in response to dietary manipulations such as a variation in energy content, dietary nitrogen composition, lipid composition or the incorporation of plant protein sources in aquafeeds (Gómez-Requeni *et al.*, 2011; Hamza *et al.*, 2010; Kolditz *et al.*, 2008; Martin *et al.*, 2003; Sveinsdóttir and Gudmundsdóttir 2010). Of particular interest in the developing finfish aquaculture industry is to understand the nutritional requirements of the faster growing and more fragile larval and juvenile stages. However, the application of the proteomic approach is more challenging during the larval stage, due to the small size of the individuals and thereby the considerable amount of material needed for such an analysis, especially when the focus is a specific tissue.

We have recently integrated two-dimensional comparative proteome analysis in studies conducted on both the larval and juvenile stages of fish. For example, some of these studies aimed to examine the impact of dietary nutrients such as protein hydrolysates or vitamin K on marine fish larvae (white seabream, Senegalese sole, gilthead seabream) metabolism, whilst other studies focused on the effect of an increased availability in dietary amino acids on liver proteome of Senegalese sole juveniles exposed to repeated handling, or muscle proteome of juveniles zebrafish. The study aiming to investigate the effects of dietary vitamin K supplementation on larvae proteome will be presented in more details.

Skeletogenesis is a critical event during larval development of fish and a high incidence of skeletal deformities is commonly observed in marine fish hatcheries, affecting growth, morphology and survival of fish, and leading to an increase in production costs and a reduction of the market value of the final products. Vitamin K is known to play an important role in bone metabolism but its positive effects over the incidence of skeletal deformities and the underlying mechanisms are still not understood. Quadruplicate groups of Senegalese sole larvae were fed from first-feeding onwards with rotifers and *Artemia* enriched with different levels of vitamin K_1 (0 to 250 mg of vitamin K_1/kg of Selco). Forty days after hatching, a comparative proteome analysis was performed by two-dimensional electrophoresis on entire larvae after removing the head and the gut. Proteins were extracted in a classical extraction buffer (containing urea, thiourea, CHAPS, dithiothreitol and a protease inhibitor cocktail), isoelectric-focused on linear gradient pH 4-7 IPG strips in the first dimension and then separated in a second dimension by SDS-PAGE on 4-12% Bis-Tris gels. Qualitative and quantitative analyses of gels were done with PDQuest v8.0 software (Bio-Rad). After spot detection and matching, the volume intensity of each spot was normalized by dividing it by the total volume intensity of valid spots. Evaluation of the statistical significance of spot variation between the two groups of larvae was performed using the non-parametric Mann-Whitney U test. Protein spots that exhibited statistically significant differences in normalised volume greater than 1.5-fold or lower than 0.67-fold between the two experimental conditions ($P<0.05$) were considered to be significantly differentially expressed. Some of the spots differentially expressed among the groups were excised from the gels, digested with trypsin and analysed by liquid chromatography-tandem mass spectrometry. Peptides mass lists generated were used for protein identification through the MASCOT search engine (Matrix Science). Two-dimensional analysis of larvae proteome allowed the detection and the comparative quantification of a total of 486 protein spots across all gels. Among these spots, 76 showed significant variations between the two experimental groups, 47 were over-expressed and 29 were under-expressed in vitamin K_1 supplemented larvae relative to control. Among over-expressed protein spots in larvae fed the vitamin K_1 supplemented diet, some were identified as being involved in clotting process (fibrinogen beta), in cellular contractile system (myosin heavy chain) and cytoskeleton proteins (cytokeratin type I). In the same experimental group of larvae, some of the under-expressed protein spots were identified as being involved in energy metabolism (enolase, creatine kinase, ATP synthase F1 complex beta polypeptide), protein folding (chaperonin containing TCP1), in cellular contractile system (myosin light chain, myosin binding protein). Vitamin K_1 supplementation also induced a down-regulation of type VI collagen, which has already been shown to be expressed in most tissues. This protein is involved in matrix structural integrity and binds other collagen fibrils and glycosaminoglycans, and is known to have a role in skeletal metabolism (Hall, 2005).

In conclusion, two-dimensional proteomics is a sensitive and valuable a tool to evaluate nutritional effects in farmed fish, including the smallest fish larvae.

Acknowledgements

NR, M de V and TSS acknowledge financial support from Fundação para a Ciência e Tecnologia, Portugal, through grants SFRH/BDP/65578/2009, SFRH/BD/40698/2007 and SFRH/BD/41392/2007 respectively. The experiments presented here were partly funded by Fundação para a Ciência e Tecnologia, through projects PDTC/MAR/105152/2008 (SPECIAL_K), PPCDT/MAR/61623/2004 (SAARGO), PTDC/MAR/71685/2006 (HYDRAA), prize CERATONIA given by Caixa Geral de Depósitos and University of Algarve, and Research Council of Norway through project 165203/S40.

References

Gómez-Requeni P., de Vareilles M., Kousoulaki K., Jordal A.-E.O., Conceição L.E.C., Rønnestad I., 2011. Whole body proteome response to a dietary lysine imbalance in zebrafish *Danio rerio*. Comparative Biochemistry and Physiology, Part D. 6:178-186.

Hall B.K., 2005. Bones and cartilage; Elsewier academic press, California USA. 760 pp.

Hamza N. Silvestre F., Mhetli M., Khemis I. B., Dieu M., Raes M., Cahu C., Kestemont P., 2010. Differential protein expression profile in the liver of pikeperch (*Sander lucioperca*) larvae fed with increasing levels of phospholipids. Comparative Biochemistry and Physiology, Part D. 5:130-137.

Kolditz C.I., Paboeuf P., Borthaire M., Esquerre´ D., Sancristobal M., Lefèvre F., Médale F., 2008. Changes induced by dietary energy intake and divergent selection for muscle fat content in rainbow trout (*Oncorhynchus mykiss*), assessed by transcriptome and proteome analysis of the liver. BMC Genomics 9:506-521.

Martin S.A.M., Vilhelmsson O., Médale F., Watt P., Kaushik S., Houlihan D.F., 2003. Proteomic sensitivity to dietary manipulations in rainbow trout. Biochim Biophys Acta 1651:17-29.

Sveinsdóttir H. and Gudmundsdóttir À., 2010. Proteome profile comparison of two differently fed groups of Atlantic cod (*Gadus morhua*) larvae. Aquaculture Nutrition 16:662-670.

Serum proteomic analysis in bovine mastitis

Paola Roncada[1] Cristian Piras[2], Luigi Bonizzi[3], Alessio Soggiu[3], Michele De Canio[4], Romana Turk[5], Mislav Kovačić[6], Marko Samardžija[7] and Andrea Urbani[4]

[1]Istituto Sperimentale Italiano L. Spallanzani, Milano, Italy; paola.roncada@guest.unimi.it

[2]DIPAV, Facoltà di Medicina Veterinaria, Università degli Studi di Milano, Italy

[3]Università degli studi di Sassari, Dipartimento di Scienze Zootecniche, Sassari, Italy

[4]Dip. di Medicina Interna, Universita' di Roma 'Tor Vergata', Roma and IRCCS-Fondazione Santa Lucia, Roma, Italy

[5]Department of Pathophysiology, Faculty of Veterinary Medicine, University of Zagreb, Zagreb, Croatia

[6]Pharmas d.o.o., Popovača, Faculty of Veterinary Medicine, University of Zagreb, Zagreb, Croatia

[7]Department of Reproduction and Clinic for Obstetrics Faculty of Veterinary Medicine, University of Zagreb, Zagreb, Croatia

Introduction

The interest in research for biomarkers discovery for the diagnoses of bovine mastitis stems largely from the need to better characterize mechanisms of the disease, to identify reliable biomarkers for use as measures of early detection and drug efficacy, and to uncover potentially novel targets for the development of alternative therapeutics. Most of proteomic studies on mastitis have been performed on milk and somatic cells[1,2]. Differential expression analysis performed from Baeker et al.[2,3] of the whey from both mastitic and non-mastitic milk revealed a marked increase in the expression of a series of some proteins during infection. Although proteomic profile of bovine milk whey proteins have been well characterized limited information has been provided on serum and plasma proteomics of bovine mastitis. The aim of this work is to extend the current knowledge on molecular circulating biomarkers of mastitis including both sub-clinical and clinical animal group. Whole serum proteome was extensively evaluated by three different complementary approaches in the clinical groups in order to possibly find differential protein expression useful to help in early diagnoses of this pathology. The collected evidences showed complementary data between oxidative stress response, lipid metabolism and the differential protein expression.

Methods

The study was conducted on a total of 80 Holstein-Frisian dairy cows located on farms in the region of Eastern Croatia. Serum protein amount was determined using Bio-Rad Protein Assay according to the manufacturer instructions. 2D electrophoresis (pI range 4-7; 12,5% AA, 7x10 cm) were carried out. After runs, gels were stained with colloidal Coomassie and digitalized with laser scanner Pharos FX (Biorad)[4]. After gel comparison with Progenesis sowtware, mass spectrometry analysis was carried out:MALDI-TOF, Shotgun analysis by nLC-MS/MS and linear MALDI. Kolmogorov-Smirnov test and Leven´s test were used for testing data for normality and equal variance. Differences between investigated groups were tested by the non-parametric analysis using Mann-Whitney rank sum-test. SigmaStat 3.0 (SPSS Inc., Chicago, Illinois, USA) was applied for statistical analysis. Statistical significance between median values was based on values $P<0.05$.

Results and discussion

Linear MALDI-TOF-MS analysis was performed in order to preliminary evaluate the sample collection quality and to possibly examine the presence of distinctive low molecular mass proteins and peptides fragments in serum samples. In fact, massive protein degradation and oxidation products have been observed whenever sample pre-analytical phase has not been properly followed[5]. Single mass spectra show very similar m/z peaks distributions. None signal indicating partial degradation of serum samples was detected. Average mass spectra

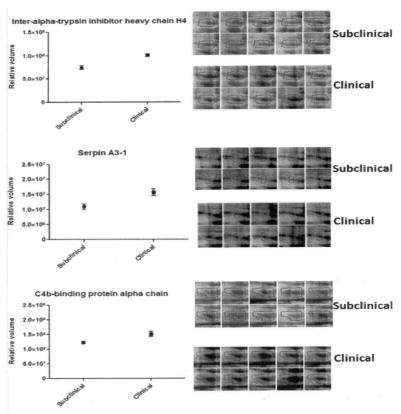

Figure 1. (left) Differentially expressed proteins in subclinical group versus control group, respectively spots no. 1,2,3 (P≤0.05).

obtained from the three class of sera (clinical mastitis, subclinical mastitis and control) were highly superimposable. Genetic Algorithm of ClinProtools software was run to construct a model for the class distribution. The algorithm failed to indicate discriminating peaks for this classification, even if recognition capability of generated model was relatively high (79.17%). This result was not unexpected indicating a substantial homogeneity in serum mass spectra. A more subtle characterisation of serum protein profile was pursued by a 2DE profiling on intact proteins. This analysis highlighted 10 differentially expressed proteins among all three groups. Three proteins were found to be differentially represented between control and subclinical group and seven proteins were found to be differentially represented between subclinical and clinical group. In the subclinical group in comparison to control group Serpine A3-1 and a vitronectin-like protein were found to be over-represented while complement factor H was found to be under-represented (Figure 1, Table 1). Seven proteins, respectively inter-alpha-trypsin inhibitor heavy chain H4, Serpine A3-1, C4b-binding protein alpha chain (Figure 2) and three different isoforms of haptoglobin were found to be upregulated in clinical group

Table 1. Protein identification.

Spot	Protein name	Trend	Accession no.	Theoretical MW(KDa)/pl	Seq. coverage	Mascot score
1	Serpin A3-1	↑ Subclinical vs. control	SPA31_BOVIN	46.3/5.6	20.7	98.20
2	Complement factor H	↓ Subclinical vs. control	CFAH_BOVIN	145.0/6.4	19.2	97.70
3	Vitronection	↑ Subclinical vs. control	Q3ZBS7_BOVIN	53.6/5.92	3.1	75
4	Inter-alpha-trypsin inhibitor heavy chain H4	↑ Clinical vs. subclinical	ITIH4_BOVIN	101.6/6.2	20.5	100.00
5	Serpin A3-1	↑ Clinical vs. subclinical	SPA31_BOVIN	46.3/5.6	29.9	114.00
6	C4b-binding protein alpha chain	↑ Clinical vs. subclinical	C4BPA_BOVIN	68.8/5.98	6.3	34
7	Haptoglobin	↑ Clinical vs. subclinical	HPT_BOVIN	45.6/8.8	28.7	79.50
8	Haptoglobin	↑ Clinical vs. subclinical	HPT_BOVIN	45.6/8.8	16.2	64.30
9	Haptoglobin	↑ Clinical vs. subclinical	HPT_BOVIN	45.6/8.8	16.5	69.10
10	Apolipoprotein A1	↓ Clinical vs. subclinical	APOA1_BOVIN	30.3/5.6	53.6	178.00

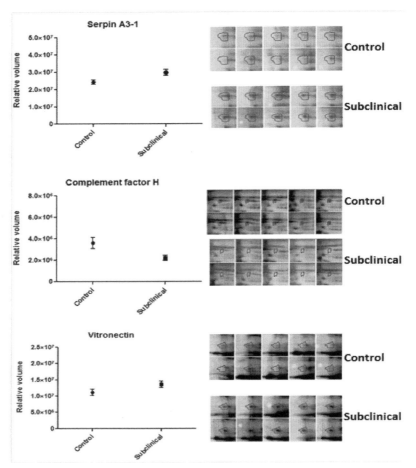

Figure 2. (right) Differentially expressed proteins in clinical group versus subclinical group, respectively spots no. 4,5,6 (P≤0.05).

in comparison to subclinical group, and only an isoform of Apolipoprotein A1 was found to be downregulated (Figure 3). All described data including mass spectrometry identification data and expression data are resumed in Table 1. Five proteins were in common between the 2DE and the nLCMS/MS shotgun experiments (data not shown).

Conclusion

Bovine mastitis is characterized by a strong release of white blood cells into the mammary gland due to bacterial invasion. Milk-secreting tissue and various ducts throughout the mammary gland are damaged due to bacterial toxins. Immune response is necessary and causes a strong increment of proinflammatory cytokines and oxidative stress that brings to higher production

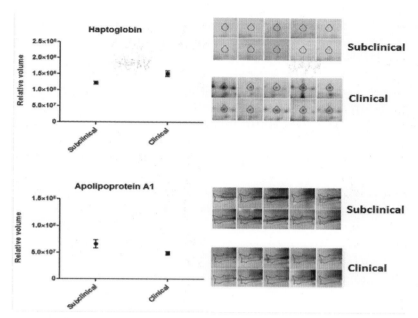

Figure 3. Differentially expressed proteins in clinical group versus subclinical group, respectively spots no. 7 and 10 (P≤0.05).

of acute phase reactants. As described we founded three differentially expressed proteins in subclinical mastitis versus control. Serpin A3-1 and vitronectin were found to be upregulated and complement factor H was found to be downregulated in subclinical mastitis. These proteins could be further explored as possible molecular biomarkers. In particular the specific vitronectin-like protein can be considered the most appropriate candidate. Its production is linked to the process of bacterial opsonization that involves both vitronectin and complement cascade. It is a serum marker that, if overexpressed, indicates the presence of a strong bacterial infection and its overexpression during subclinical mastitis could help in the early diagnoses of this pathology. During the clinical mastitis in comparison to subclinical mastitis it was found the overexpression of three different isoforms of haptoglobin and apoliporotein A1. The possibility of the interaction of these two proteins was already described[6]. In particular the interaction and the lower renal excretion of these two proteins could be due to the formation of a complex between apoA1 and haptoglobin due to oxidative stress that is enhanced during inflammation.

References

1. Boehmer JL. Proteomic Analyses of Host and Pathogen Responses during Bovine Mastitis. Journal of Mammary Gland Biology and Neoplasia. 2011:1-16.

2. Alonso-Fauste I, Andrés M, Iturralde M, Lampreave F, Gallart J, Álava MA. Proteomic characterization by 2-DE in bovine serum and whey from healthy and mastitis affected farm animals. Journal of proteomics. 2011 Dec 13. [Epub ahead of print]

3. Baeker R, Haebel S, Schlatterer K, Schlatterer B. Lipocalin-type prostaglandin D synthase in milk: a new biomarker for bovine mastitis. Prostaglandins &Other Lipid Mediators. 2002;67:75-88.

4. Roncada P, Begni B, Amadori M, Cristoni S, Archetti I, Boldetti C, *et al.* Blood Serum Proteome for Welfare Evaluation in Pigs. Vet Res Commun. 2007;31:321-5

5. Pieragostino D, Petrucci F, Del Boccio P, Mantini D, Lugaresi A, Tiberio S, *et al.* Pre-analytical factors in clinical proteomics investigations: Impact of ex vivo protein modifications for multiple sclerosis biomarker discovery. Journal of proteomics. 2010;73:579-92.

6. Santucci L, Candiano G, Petretto A, Pavone B, Bruschi M, Gusmano R, *et al.* Protein–protein interaction heterogeneity of plasma apolipoprotein A1 in nephrotic syndrome. Mol BioSyst. 2010.

From protein markers to phenotyping tools for evaluation of beef tenderness

Brigitte Picard, Bruno Meunier, Christiane Barboiron, Nicole Dunoyer, Nicolas Guillemin and Didier Micol
INRA UMRH 1213, Theix, 63122 Saint-Genès-Champanelle, France;
brigitte.picard@clermont.inra.fr

High and uncontrolled beef tenderness variability is at the origin of a dissatisfaction of the beef consumers. Today tenderness can be estimated only after slaughter, by sensory analysis tests and/or mechanical measurements. For beef producers, it is of interest for breeding or finishing purposes to predict the ability of live animals to produce good meat, with specific attention towards tenderness. Thus, the Beef sector is looking for biological or molecular indicators that would identify live animals with desirable quality attributes, in order to direct them towards the most appropriate production system. Tenderness is a complex trait under multifactorial determinant leading to a difficult control. For several years, various genomics programs have been conducted at the national and international level in order to reveal genomic markers of tenderness (Cassar-Malek *et al.*, 2008). Comparative transcriptomic and proteomic analysis, conducted on bovine muscles with low or high tenderness scores estimated by sensory analysis and/or mechanical measurements, bring up a list of potential biological markers which may be used as phenotypic markers to predict the 'tenderness potential' of an animal or a carcass. This phenotypic analysis could be done either at the mRNA or protein level. In this abstract, we will focus on the protein level. The strategy developed by our team over a few years is: 1) to validate the relationship between meat tenderness and the protein markers on a large number of cattle types as the most representative of the French production systems and beef consumption (male and female, beef and dairy breeds); 2) to determine the effects of management factors (age and diet) on the expression of these markers; 3) to develop an antibody micro-array prototype for molecular phenotyping of beef tenderness from the list of validated markers. This antibody micro-array will be available for the Beef industry for evaluation of meat quality and for optimization of cattle breeding for meat production. This tool could also be used for tenderness phenotyping in genomic breeding schemes.

Material and methods

The results presented in this abstract come from several experiments of our team. The muscle studied was the *Longissimus thoracis* (LT) sampled on young bulls from different breeds: Blonde d'Aquitaine, Charolaise, Limousine, Salers, Angus with a mean age of 17 months. The LT muscle was taken just after slaughter or after 24h, frozen directly in liquid nitrogen and stored at -80 °C until use. On these LT samples, tenderness was evaluated by sensory analysis and/or mechanical measurement (Warner-Bratzler test with INSTRON 4411) after 14 days ageing in vacuum packaging and cooking in a grill until an internal temperature of 55 °C.

For proteomic analyses, samples were chosen on the basis of their low or high tenderness scores estimated by sensory analysis and/or mechanical measurement. Proteins were extracted from LT muscle and separated by two-dimensional electrophoresis (2DE) according to the technique described by Bouley *et al.* (2004) in a 4-7 pH gradient. After image analysis (Same Spots) of the gels from the two groups of tenderness (n=5 to 10 samples/group), proteins with significantly different level of abundance between groups have been detected (Meunier *et al.*, 2005). The spots were extracted from acrylamide gel and digested with trypsin prior to their identification by mass spectrometry Maldi-Tof or MS/MS (Bouley *et al.*, 2004; 2005) at the INRA Plateform 'Metabolism functional exploration: from genes to metabolites' (PEFM).

This comparative proteomic analysis gave a list of protein identified as potential markers of beef tenderness (for review see: Picard *et al.*, 2010). In order to validate the relationship between the abundance of these proteins and tenderness on a large number of samples, we have developed a dot-blot assay. Dot-blot is a useful immunological technique which allows the quantification of proteins using specific antibodies on a large number of samples simultaneously (Guillemin *et al.*, 2009). In a first step, the specificity of the antibodies and their conditions of use on bovine muscle, have to be validated by western-blot. For dot-blot analysis, the protein extractions prepared for 2DE were spotted on a nitrocellulose membrane, incubated with the primary specific antibody, and then with a fluorescent secondary antibody as described by Guillemin *et al.* (2009).

Results and discussion

Comparative proteomic analysis from different experiments of our team gave a list of 24 proteins identified as differently abundant between extreme groups of tenderness (for review see Picard *et al.*, 2010). The toughest LT muscles have higher abundance of proteins representative of the fast glycolytic type. This is the case of phosphoglucomutase (PGM), lactate dehydrogenase B, triosphosphate isomerase, glyceraldehyde 3-phosphate dehydrogenase and fast isoforms of troponin T in Charolais and Blonde d'Aquitaine breeds. In contrast, proteins of slow oxidative type are more abundant in the most tender meat. These include the α Enolase, an isoform of troponin T slow, slow isoforms of myosin light chains, the creatine kinase M and the mitochondrial protein NADH-ubiquinone oxidoreductase. Other experiments of the literature lead to similar conclusions about the relationship between oxidative metabolism and tenderness in LT muscle (Picard *et al.*, 2010). Moreover, the results of Bouley *et al.* (2005) showed that some proteins such as isoforms of fast troponin T, the Myosin Binding Protein-H and PGM appear correlated with both muscle mass and meat tenderness. This shows that it is possible to control both the amount of meat produced and its tenderness.

Several proteins involved in calcium metabolism have been identified as positive marker of tenderness (Picard *et al.*, 2010). For example, paravalbumin is more abundant in the group of high tenderness in Charolais and Limousine breed. This protein has structural properties that give it a high affinity for calcium ions. It participates, particularly in fast fibres in calcium cycle between the cytoplasm and the sarcoplasmic reticulum. On the other hand, the myosin light chain 2 whose abundance is lower in the group of high tenderness in the Charolaise and Limousine beef breeds has a calcium binding site. In the same way, the abundance of acyl-CoA-binding protein is increased in the higher group of tenderness especially in Limousine breed. One of the roles of the complex Acyl-coA/ACBP is to regulate the release of calcium ions from the sarcoplasmic reticulum by increasing the activity of calcium channels. So, the calcium cycle proteins appear to be strongly involved in meat tenderness.

Structural proteins such as actin and α CapZβ (binding protein actin) appear to discriminate between groups of sensory or mechanical tenderness in several beef breeds.

All our studies revealed an important role of heat shock proteins (Hsp). For example, a negative relationship between the expression of DNAJA1 and tenderness was demonstrated in LT muscle of Charolais cattle (Bernard *et al.*, 2007). This gene encodes Hsp40 protein involved in the entry of proteins into mitochondria and inhibiting apoptosis in the interactive mechanism with another (Hsp70) chaperone protein. This anti-apoptotic activity could slow down the process of cell death in the early stages of meat ageing. Other proteins of the same family have been identified as markers of tenderness in several experiments. As example Hsp27 protein was found correlated with tenderness in several independent studies (Picard *et al.*, 2010). The Hsp27 protein interacts with other partners, particularly Hsp70 family and α-crystallin B with which it forms a dynamic complex that protects particularly the enzymes of oxidative metabolism and myofibrillar proteins (as desmin, actin, myosin, titin) during cellular stress. In the live animal, the large amount of Hsp27 protects the actin degradation. However after slaughter, Hsp27 prevents protein aggregation and promote the access of proteases to their targets increasing proteolysis of actin. The protein Hsp27 via its anti-apoptotic role is therefore a good candidate to be a relevant tenderness marker. Our results revealed also a role in tenderness of proteins involved in oxidative stress such as Super-oxide dismutase and Peroxiredoxin 6.

From this list of protein markers of beef tenderness, we have developed a bioinformatic analysis in order to have a better understanding of the biological functions involved in tenderness and the interactions between all these proteins (Guillemin *et al.*, 2011). Currently we are currently developing an antibody micro-array by using the antibodies corresponding to these protein markers (Sakanyan and Angelini, 2009). This tool will be used to quantify the abundance of all the protein markers simultaneously for a muscle sample.

Conclusion

At the scientific level, this approach allowed us to give new insights about the biological functions involved in meat tenderness. Moreover, it provided the development of a high-throughput screening method to quantify the abundance of tenderness markers at the protein level as Dot-Blot and antibody-microarray. This last tool will be used for a 'paddock' application. This technology is already used in Medical Science as a diagnostic tool. It will be used on muscle samples taken by biopsies on live animal or on carcass after slaughter. This will be internationally the first phenotyping tool for meat tenderness.

References

1. Bernard C., Cassar-Malek I., Le Cunff M., Dubroeucq H., Renand G., Hocquette J.F. 2007. New indicators of beef sensory quality revealed by expression of specific genes. Journal of Agricultural and Food Chemistry, 55(13):5229-5237.
2. Bouley J., Chambon C., Picard B. 2004. Mapping of bovine skeletal muscle proteins using two-dimensional gel electrophoresis and mass spectrometry. Proteomics, 4 (6): 1811-1824.
3. Bouley J., Meunier B., Chambon C., De Smet S., Hocquette J.F., Picard B. 2005. Proteomic analysis of bovine skeletal muscle hypertrophy. Proteomics, 5: 490-500.
4. Cassar-Malek I., Picard B., Bernard C., Hocquette J.F. 2008. Application of gene expression studies in livestock production systems: a European perspective. Australian Journal of Experimental Agriculture, 48: 701-710.
5. Guillemin N, Meunier B, Jurie C, Cassar-Malek I, Hocquette JF, Levéziel H and Picard B 2009. Validation of a Dot-Blot quantitative technique for large-scale analysis of beef tenderness biomarkers. Journal of Physiology and Pharmacology, 60: 91-97.
6. Guillemin N., Bonnet M., Jurie C., Picard B. 2011. Functional analysis of beef tenderness. Journal of proteomics, Journal of Proteomics, 75: 352-365.
7. Meunier B., Bouley J., Piec I., Bernard C., Picard B., Hocquette J.F. 2005. Data analysis methods for detection of differential protein expression in two-dimensional gel electrophoresis. Analytical Biochemistry, 340: 2, 226-230.
8. Picard B., Berri C., Lefaucheur L., Molette C., Sayd T. Terlouw C. 2010. Skeletal muscle proteomics in livestock production. Briefings in Functional Genomics and Proteomics, 9: 259-278.
9. Sakanyan V., Angelini M. 2009. Dispositifs de puces à molécules et leurs utilisations. Brevet. PCT/EP2009/050635.

Immunoreactive proteins of *Mycobacterium avium* subsp. *paratuberculosis*

C. Piras[1], A. Soggiu[2], L. Bonizzi[2], A. Urbani[3,4], V. Greco[3,4], G.F. Greppi[1], N. Arrigoni[5] and P. Roncada[6]

[1]Dipartimento di Scienze Zootecniche, Centro NanoBiotecnologie, Università degli studi di Sassari, Sassari, Italy; cristian.piras@uniss.it

[2]Dipartimento di Patologia Animale, Igiene e Sanità Pubblica Veterinaria, Facoltà di Medicina Veterinaria, Università Degli Studi di Milano, Milano, Italy

[3]Dipartimento di medicina interna,Università Tor Vergata, Roma, Italy

[4]Fondazione Santa Lucia – IRCCS, Rome, Italy

[5]Sezione Diagnostica di Piacenza, Centro di Referenza Nazionale per la Paratubercolosi, IZSLER, Piacenza, Italy

[6]Istituto Sperimentale Italiano L. Spallanzani, Milano, Italy

Introduction

Johne's disease is a *Mycobacterium avium* subsp. *paratuberculosis* (MAP)-caused chronic enteritis of ruminants (bovine paratuberculosis) associated with enormous worldwide economic losses for the dairy products industries. MAP is a slow growing bacterium that can infect ruminants and remain latent for years without development of any clinical signs disease evident symptoms. The actual diagnostic tools are able to diagnose this pathology only after years of infection when a relevant number of other animals has been infected. Spread of this pathology causes economic losses because of reduced milk yield, premature culling and reduced slaughter value[1].

Eradication programs and the management limitations for this disease have been hampered by the lack of simple and specific diagnostic tests for detecting the disease in subclinically infected (infected but symptom-free) animals. Diagnosis is based on detection of antibodies in milk or serum, or by bacterial culture from faeces, but these diagnostic methods are usually applicable only after years of infection when the disease is already in an advanced status[2].

A good and sensitive diagnostic method is required to avoid the spread end the eradication of this pathology that afflicts most ruminant species of farm animals. For this reason an immunoproteomic approach represents a good chance to investigate new biomarkers in cattle paratuberculosis diagnoses.

Methods

Mycobacterial lysis was performed using both freeze-thaw cycles and bead beating method. MAP proteins were immunoblotted respectively against control and MAP infected sera.

Briefly, about 95 mg of pelleted MAP cells were diluted in 750 µl of a buffer containing 7M urea, 2M thiourea, 4% CHAPS, 2% ampholines, 0,8% DTT and protease inibitors. After a bead beating cycle of 30 minutes (400 µg of Biospec zirconium-silica beads/750 µl buffer) the obtained lysate was quickly frozen at -80 °C in a metal block. After 2 h at -80 °C the sample was quickly defrost in a metal block. This procedure was repeated three times. After the last cycle the sample was centrifuged at 9,000x*g* for 10 minutes to remove zirconium-silica beads and cell debris. The supernatant was stored at -80 °C until use.

80 µg of protein was loaded onto 4-7 pH gradient home-made IPG strips. After IEF the second dimension was performed on 12% acrylamide gels. Obtained gels were electroblotted on PVDF membranes for 1h on a semidry immunoblotting apparatus.

Membranes with MAP proteins were immunoblotted against eight different animal sera, four sera belonged from healthy control animals and four from paratuberculosis affected animals. Immunogenic proteins were founded by subtraction of immunoreactive spots present in the control group maps from the spots present in paratuberculosis group maps. The spots of interest were excised from the gel and analyzed through mass spectrometry.

Results

As results were founded after 2D immunoblotting image analysis (Figure 1) 11 immunoreactive spots present only when MAP proteins were immunoblotted against paratuberculosis infected serum (Figure 2) most of them were successfully identified with mass spectrometry as shown in Table 1.

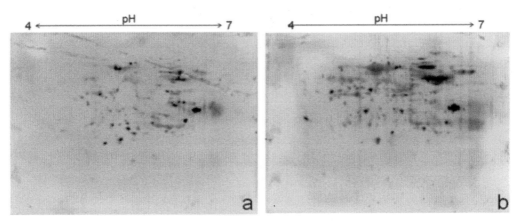

Figure 1. 2D immunoblotting of MAP electroblotted proteins against control bovine serum (a) and MAP infected bovine serum (b).

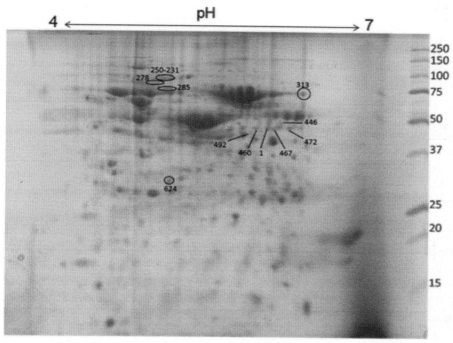

Figure 2. Immunoreactive spots identified in MAP maps and analyzed through mass spectrometry.

Table 1. Protein names of identified immunoreactive proteins.

Spot number	Protein
1	Soluble pyridine nucleotide transhydrogenase
250	Pd2U protein (*Mycobacterium avium* 104)
278	Phosphoglucosamine mutase (*Mycobacterium avium* subsp. *paratuberculosis*)
313	Transcriptional regulator, TetR family protein (*Mycobacterium avium* subsp. *paratuberculosis*)
446	Translation initiation factor IF-1 (*Mycobacterium avium* subsp. *paratuberculosis*)
460	Ianosterol 14-alpha demethylase
467	3-ketoacyl-ACP reductase (*Mycobacterium avium* subsp. *paratuberculosis*)
472	RNA polymerase sigma factor RpoD, C-terminal domain/RNA polymerase sigma factor, sigma-70 family (*Mycobacterium avium* subsp. *paratuberculosis*)
492	Hypothetical protein MAP0323c (*Mycobacterium avium* subsp. *paratuberculosis*)
624	ATP-dependent protease ATP-binding subunit (*Mycobacterium avium* subsp. *paratuberculosis*)

Discussion

In this work was developed a powerful extraction and solubilization method for MAP proteins by coupling freeze thaw cycles with bead beating. This method is able to give a high protein recovery rate, a good rate of MAP cell disruption, nonetheless a clean sample that does not require any precipitation procedure before IEF.

Immunoreactive proteins identified are putative molecular targets that, after validation of their specificity, could be used as immunoreactive targets for serological tests, for the production of recombinant antigens[3] or for developing antibodies for MAP specific detection.

References

1. Ott, S.L., Wells, S.J. & Wagner, B.A. Herd-level economic losses associated with Johne's disease on US dairy operations. Preventive Veterinary Medicine 40, 179-192 (1999).
2. Nielsen, S.S. & Toft, N. Age-specific characteristics of ELISA and fecal culture for purpose-specific testing for paratuberculosis. Journal of dairy science 89, 569-579 (2006).
3. Mikkelsen, H., Aagaard, C., Nielsen, S.S. & Jungersen, G. Review of Mycobacterium avium subsp. paratuberculosis antigen candidates with diagnostic potential. Veterinary Microbiology 152, 1-20 (2011).

Acute phase protein expression in different visceral and subcutaneous fat depots from clinically healthy dairy cows

Md. Mizanur Rahman[1], Susanne Häussler[1], Manfred Mielenz[1], Cristina Lecchi[2], Fabrizio Ceciliani[2] and Helga Sauerwein[1]

[1]Institute of Animal Science, Physiology & Hygiene, University of Bonn, 53115 Bonn, Germany; sauerwein@uni-bonn.de

[2]Department of Animal Pathology, Hygiene and Veterinary Public Health, University of Milan, Italy

Introduction

White adipose tissue (AT) has been considered for a long time as the store house of energy and this tissue is now recognized as an active endocrine organ that communicates with the brain and peripheral tissues by secreting a wide range of hormones and cytokines collectively termed adipokines; the adipokine family includes also several acute phase proteins (APP) (1). APP are part of the innate immunity and interplay in the restoration of homeostasis and the restraint of microbial growth before acquired immunity is active. Measurements of APP responses from physiological to pathological changes also help in monitoring and diagnosis of diseases in animals (2,3). Liver is considered as the principal organ of APP production, but extra hepatic production of alpha1-acid glycoprotein (AGP), serum amyloid A (SAA), haptoglobin (Hp), and lipopolysaccharide binding protein (LBP) has also been reported (4-9). APP production from AT has been reported for humans and rodents. Recently, SAA mRNA expression (13) from bovine adipose tissue has been confirmed by our group. However, information of APP production from different fat depots is not available at the protein level according to our knowledge. Investigation of APP in fat depots in physiological condition will improve our understanding about the association of AT and APP.

Materials and methods

Adipose tissue from different fat depots namely omental, mesenteric, retroperitoneal and tail head was collected at 105 days postpartum from a clinically healthy control cow within the frame of a feeding study (14). Immediately after dissection, the samples were kept in liquid nitrogen and preserved at -80 °C for future use. Gradient PCR was performed to check the qualitative AGP mRNA expression in retroperitoneal and tail head fat and in liver tissue. The amplification of the coding sequence was performed in Alphase® Thermal cycler using AGP primers (Accession Nr. AM403243) (Forward 5'-GCATAGGCATCCAGGAATCA-3' and Reverse 5'-TAGGACGCTTCTGTCTCC-3'). The thermal profile was: 2 min 72 °C; 1 min 95 °C; 40 sec 94 °C; 40 sec 60 °C and 55 °C, and 40 sec-72 °C, total 40 cycles.

Aliquots of liver and fat tissue were homogenized for Western Blot analyses using HEPES buffer (10 mM pH 7.4) in the Precellys® system. Total protein was quantified by the Bradford method. An appropriate amount of total protein depending upon the specific APP was separated by sodium dodecyl sulphate-polyacrylamide gel electrophoresis (SDS–PAGE) and Western blotted onto polyvinyl difluoride (PVDF) membranes. The membranes were immunolabelled for the presence of APPs using specific antibodies (Table 1) and immunoreactive bands were visualised by enhanced chemiluminescence (ECL). To confirm that an equal amount of protein was loaded on each lane, membranes were stripped and immunolabelled with a mouse anti β-actin antibody (Table 1). Recombinant human LBP (Biometec Ltd, Greifswald,Germany), purified bovine AGP and bovine serum were used as positive controls in LBP, AGP and SAA Western Blot analysis.

Table1. APP antibodies used for western blot analyses.

Primary antibody	Dilution	Incubation time (h)	Ref.	Sample loaded	Gel (%)
Rabbit anti-bov AGP	1:2,000	0.45	10	Fat/Liver-5µg AGP-75ng	12
Mouse anti-humanLBP	1:600	Overnight/16	9	Fat/Liver-5µg LBP-15ng	12
Rabbit anti-bov SAA	1:25,000	1.30	7	Fat-25/50 µg	15
Mouse anti-human β-actin	1:2,000	Overnight/16		---------	-----

Results

Initially, AGP mRNA expression was confirmed both in fat depots and in liver by qualitative PCR (Figure 1).

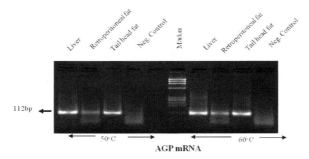

Figure 1. AGP mRNA expression in retroperitoneal and tail head fat depots and in liver.

AGP, LBP and SAA were detected by Western Blot analysis. AGP was expressed as 45 kDa and 66 kDa bands in different fat depots which are presumably glycosylated forms. There were also faint bands at 36 kDa and 23 kDa. Bovine deglycosylated AGP is about 20 kDa (6). Liver and bovine serum also showed similar band patterns but the bands were weaker in comparison to the ones obtained from fat. LBP immunoreactive bands were obtained at 55 and 67 kDa. The original protein of 50 kDa was observed in liver (Figure 2), but other bands were also obtained thus yielding different patterns of bands in liver as compared to fat and to serum. SAA was detectable as weak band of 13 kDa in mastitic bovine serum (positive control); in mesenteric fat and in bovine serum additional bands presumably representing the dimer (26 kDa) and the tetramer form of SAA were also observed (Figure 2).

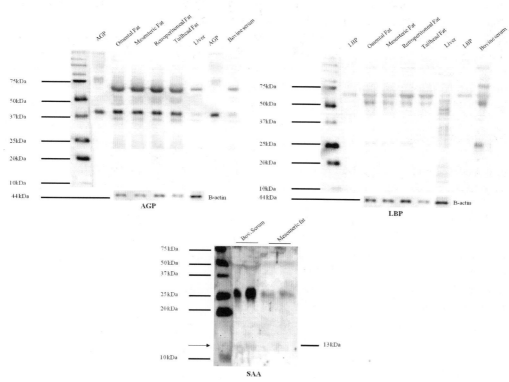

Figure 2. SDS–PAGE and Western blot analysis of different fat depots and liver. Acute phase proteins were identified after immunostaining using antibodies listed in Table 1 and detection by ECL.

Discussion and conclusion

APP are important in the preliminary innate immune response of the host to pathogens, trauma and other noxa. AGP is mainly a binding protein with immunomodulatory properties which is expressed at high levels in liver in comparison to several other organs in cattle (10,5). In contrast, our data indicate that different fat depots in cattle can produce higher amounts of the AGP protein than liver. The role of LBP is to mount immune response against invading bacteria in the host through detoxification or removal of bacterial toxin and bacteria (11,12). Our results support that in addition to other LBP sources in the body; fat depots can also contribute to this homeostatic reaction. SAA is a major APP in cattle with immunomodulatory functions including opsonization of bacteria (7). Immuno-reactive bands of SAA in adipose tissue were revealed as dimer and tetramer. In bovine serum the monomeric band was weak and most of the protein loaded seemed to be accumulated as dimer. Even in fat we did not find any signal at the position corresponding to the monomer. It is possible that on long time storage and repeated freezing and thawing, SAA forms multimers (personal communication, A. Molenaar, 2011). Finally we can conclude that like many other extra hepatic sources, adipose tissue from different depot localizations in cattle can produce APP which may contribute in local or systemic innate immune response.

Acknowledgements

The award of a research stipendium of the Alexander von Humboldt Foundation, Germany to Md. M. Rahman is gratefully acknowledged.

References

1. Trayhurn *et al.*, 2008. Arch Physiol Biochem, 114(4):267-76.
2. Gruys *et al.*, 1994. Vet. Bulletin 64, 1009-1018.
3. Eckersall, 2000. Revue de Medicine Veterinaire 151, 577-584.
4. Ceciliani *et al.*, 2005. Vet Res. 36(5-6): 735-46.
5. Lecchi *et al.*, 2009. Vet J. 180(2):256-8.
6. Rahman *et al.*, 2008. Vet Immunol Immunopathol. 125(1-2):71-81.
7. Molenaar *et al.*, 2009. Biomarkers. 14(1):26-37.
8. Thielen *et al.*, 2007. J Dairy Sci. 90(3):1215-9.
9. Rahman *et al.*, 2010. Vet Immunol Immunopathol. 137 (1-2): 28-35.
10. Ceciliani *et al.*, 2007. Vet. Immunol Immunopath, 116, 145-152.
11. Zweigner *et al.*, 2006. Microbes Infect. 8 (3), 946-952.
12. Schroder *et al.*, 2003. J. Biol. Chem. 278 (18), 15587- 15594.
13. Saremi *et al.*, 2011. J. Dairy Sci. Vol. 94 (E-Suppl. 1): 596
14. Von Soosten *et al.*, 2011 J Dairy Sci 94: 2859-2870

The plasma proteome and acute phase proteins of broiler chickens with gait abnormalities

E.L. O'Reilly[1], R.J. Burchmore[2], V. Sandilands[3], N.H. Sparks[3], C. Walls[1] and P.D. Eckersall[1]
[1]Institute of Infection, Inflammation and Immunity, College of Medicine, Veterinary and Life Sciences, Glasgow University, Glasgow, Scotland, United Kingdom; e.o'reilly.1@research.gla.ac.uk
[2]Glasgow Polyomics Facility, College of Medicine, Veterinary and Life Sciences, Glasgow University, Glasgow, Scotland, United Kingdom
[3]Avian Science Research Centre, Scottish Agricultural College, Auchincruive, Ayr, Scotland, United Kingdom

The acute phase response (APR) is the early and non-specific systemic reaction of the innate immune system to homeostatic disturbances. Pro-inflammatory cytokines and chemokines released from macrophages, monocytes and infected and damaged tissues affect the synthesis and secretion of hepatocytic proteins and drastically alter the plasma protein profile. Plasma proteins that change concentrations as a result of an APR are termed acute phase proteins (APPs). The measurable changes in APP concentrations during infectious, inflammatory, stressful, traumatic or neoplastic events are often proportional to the severity of the event(s). As such APP measurements are used as disease biomarkers and for prognostication.

Gait abnormalities in broiler (meat) chickens and the diseases and disorders that cause them are major issues for the poultry industry affecting both productivity and welfare. As such there is a need to investigate alternative methods to identify, prevent and control the underlying causes of gait abnormalities in broilers.

This investigation characterised APP changes that occur due to gait abnormalities in broilers by measurement of the established APPs in chickens: ceruloplasmin (Cp), PIT54 (an avian haemoglobin binding protein similar to mammalian haptoglobin) and ovotransferrin (OVT) which are known to increase in response to bacterial, viral and parasitic infection (Rath *et al.* 2009; Georgieva, *et al.* 2010). Furthermore the acute phase reactive plasma proteome was determined with the view of identifying new biomarkers and biomarker profiles. Few studies have investigated the 2D proteome of chickens, though changes in the serum protein proteome due to layer hen age and coccidial infection have been demonstrated (Huang *et al.*, 2006; Gilbert *et al.*, 2011).

Methods

43 male Ross and Cobb broiler chickens aged 35-36 days were selected from various commercial farms on the basis of having a gait score of 1 (GS1), a slight uneven gait typical of normal broiler walking, or GS3, an obvious gait defect affecting ability to move about (Kestin *et al.* 1992). Birds were weighed and blood for plasma preparation recovered. A Pentra biochemical

auto analyser was used to measure Cp indirectly using p-phenylenediamine oxidase activity (Ceron *et al.* 2005) and PIT54 concentration was calculated based on the haemoglobin binding activity (Eckersall *et al.* 1999). OVT was measured using an indirect competitive enzyme immunoassay (Rath *et al.* 2009). Data was analysed as general linear mixed models with the 'R' statistical program v2.12.1.

For analysis of the acute phase reactive plasma (APRP) proteome of the GS3 birds, samples were ranked according to plasma concentrations of Cp, PIT54 and OVT and the plasma samples (n=3) with the highest across all APPs selected. APP results from the GS1 birds were also ranked and samples with the lowest APP concentrations selected (n=3) for further proteomic investigation (non-APRP). The plasma proteome was analysed using 2D SDS PAGE. For the focusing, 200 μg of plasma protein diluted in rehydration buffer was loaded onto 11 cm 3-10 pH non-linear IPG strips (ReadyStripTM BioRad, Hemel Hemstead, UK) and focused overnight. For the second dimension IPG strips were reduced then acylated in equilibration buffer and inserted into the well of a IPG+1 TGX precast 4-12% acrylamide gel (Criterion, BioRad). The gel was run for 50 minutes in MOPS buffer at 200V. Gels were stained in Coomaissie blue. Protein spots were excised from the gels and analysed on Bruker AmaZon ion trap mass spectrometry with comparisons to the MASCOT protein database.

Results and discussion

Plasma concentrations (Table 1) of all three APPs were significantly associated with gait, with GS3 birds having significantly higher Cp ($P \leq 0.001$), PIT54 ($P \leq 0.05$) and OVT ($P \leq 0.05$) than GS1 birds, indicating that increased APP expression is associated with poor GS in broilers. Within the groups there was considerable variation in the APP concentrations which may indicate that sub-populations exist related to differing aetiologies of gait defect.

Cp was significantly associated with weight with heavier birds tending to have higher Cp concentrations ($P \leq 0.05$). Although no significant difference was identified between weight and gait score, higher Cp in the heavier birds may indicate that increasing weight results in

Table 1. APP ranges and medians in GS1 and GS3 birds.

	Gate score 1			Gait score 3		
	lowest	highest	median	lowest	highest	median
Cp g/l	0.275	2.352	0.698	0.037	4.440	0.815
PIT54 g/l	0.008	0.186	0.105	0.059	0.188	0.115
OVT g/l	0.750	3.668	2.210	0.413	11.903	2.065

a series of events, inflammatory in nature, that result in higher plasma Cp concentrations. There were no significant associations between the APP concentrations and broiler breed. APP concentration was used to select plasma from the most acute phase reactive GS3 birds and compared to non-APRP from GS1 birds. Differences were found in a number of regions on the APRP and non-APRP gels, with many spots differing in presence and intensity. The 3 non-APRP samples gave consistent plasma proteomes (Figure1a).There was more variation in the proteomes of the 3 APRP samples, with additional spots appearing and apparent differences in the intensity of spots (Figure 1b).

Mass spectrometry and peptide fingerprinting identified proteins illustrated in Figure 2. Established APPs OVT and PIT54 were found, however Cp was not identified on the 2D gel. Other APPs identified included haemopexin, α-1-acid glycoprotein and fibrinogen. In contrasting the plasma proteomes of APRP from the GS3 birds to the non-APRP of control birds areas corresponding to immunoglobulin (Ig) have increased while the OVT spots are of a higher intensity. The α-1-acid glycoprotein spot was well defined in the APRP gels and was either faint or completely absent in the non-APRP gels. Many of the proteins identified were found previously in the chicken proteome by Huang *et al.* (2006) and Gilbert *et al.*, (2011), however this is the first report of the acute phase proteome of chickens with gait abnormalities from commercial rearing farms. Further proteomic work, which may benefit from the removal of albumin and immunoglobulin, is required to further investigate new biomarkers and establish biomarker profiles including a routine chicken APP panel. These proteomic developments may help in addressing leg weakness, its underlying causes and other conditions that impact negatively on bird welfare and cause significant economic losses to the poultry industry.

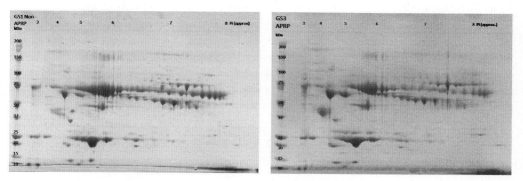

Figure 1. 2D gels comparing non-acute phase reactive and acute phase reactive plasma. a: Non-APRP plasma; b: APRP plasma.

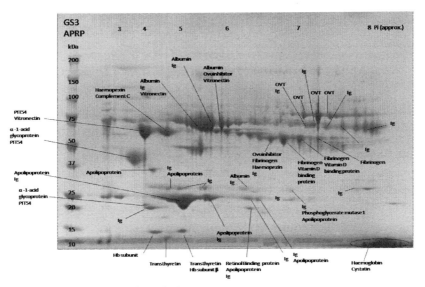

Figure 2. Summary of proteins identified using mass spectrometry.

Acknowledgements

The BBSRC and DEFRA are thanked for support of this project.

References

Ceron, J J; Eckersall, P D; Martinez-Subiela, S (2005) acute phase proteins in dogs and cats: current knowledge and future perspectives *Veterinary Clinical Pathology* 34: 85-99

Eckersall, P D; Duthie, S; Safi, S; Moffatt, D; Horadagoda, N U; Doyle, S; Parton, R; Bennett, D; Fitzpatrick, J L (1999) An automated biochemical assay for haptoglobin: prevention of interference from albumin *Comparative Haematology International* 9: 117-124

Georgieva, T M (2010) Similarities and differences between haptoglobin in mammals and PIT54 in poultry – a review *Bulgarian Journal of Veterinary Medicine* 13: 1-12

Gilbert, E. R., Cox, C. M., Williams, P. M., McElroy, A. P., Dalloul, R. A., Ray, K., Barri, A., Emmerson, D. A., Wong, E, A., Webb, K. E. (2011) Eimeria species and genetic background influences the serum protein profile of broilers with Coccidiosis *PLOS ONE* 6: e14636

Huang, S., Lin, J., Chen, Y., Chuang, C,. Chiu, Y., Chen, M., Chen, H., Lee, W. (2006) Analysis of chicken serum proteome and differential protein expression during development in single comb white leghorn hens *Proteomics* 2: 2217-2224

Kestin, S.C, Knowles, T.G., Tinch, A.E and Gregory, N.G. (1992) Prevalence of leg weakness in broiler chickens and its relationship with genotype *Veterinary Record*, 131, 190-196

Rath, N C; Anthony, N B; Kannan, L; Huff, W E; Huff G R; Chapman, H D; Erf, G F; Wakenell, P (2009) Serum ovotransferrin as a biomarker of inflammatory diseases in chickens *Poultry Science* 88: 2069-2074

Proteomic analysis of visceral and subcutaneous adipose tissue in goats

Laura Restelli[1], Dorte Thomassen[2], Marius Cosmin Codrea[2], Giovanni Savoini[3], Fabrizio Ceciliani[1] and Emoke Bendixen[2]

[1]*Department of Animal Pathology, Hygiene and Public Health; Faculty of Veterinary Medicine, Università degli Studi di Milano, Milan, Italy; laura.restelli@unimi.it*

[2]*Department of Animal Science; Faculty of Science and Technology, Aarhus Universitet, Tjele, Denmark*

[3]*Department of Animal Science and Technology for Food Safety; Faculty of Veterinary Medicine, Università degli Studi di Milano, Milan, Italy*

Introduction

Adipose tissue (AT) has for a long time been regarded as a simple storage tissue for energy in the form of triglycerides. There is currently accumulating scientific evidence that adipose tissue is a highly active organ, that actively takes part in regulating several metabolic processes, including reproduction, inflammatory response, and production and secretion of signalling molecules that have important biological roles, also known as adipokines. These most importantly include tumor necrosis factor alpha (TNF-alpha), interleukins, leptin and adiponectin.

Adipose tissue should not be considered a single endocrine organ, but a group of endocrine organs each of them showing depot-specific endocrine function (Kershaw and Flier, 2004). Studies regarding adipose tissue endocrine function are mainly focused on subcutaneous fat but Lemor *et al.* demonstrated that different depots of visceral and subcutaneous origin differ in mRNA abundance of several adipokines underlying the importance of sampling site when the aim is to characterize the metabolic role of AT (Lemor *et al.*, 2010).

The metabolic role of AT is very important in animal husbandry. However, most of the studies regarding this field have been done in humans and rodents.

To the best of our knowledge, only one study has been carried out on proteomic bovine adipose tissue differences so far (Rajesh *et al.*, 2010), and none in goats. Likewise, the influence of diets with different fatty acids composition on the development of fatty tissues have not yet been investigated at the proteome level.

This project was carried out at Aarhus University (Denmark) by means of a COST-Short Term Scientific Mission grant and it is part of a larger one aimed to elucidate the impact of diets enriched with different type of fatty acids on goats peripartum, including the deposition of adipose tissue in newborns.

Aim of the work

The aim of this project was to perform proteome characterizations of different adipose tissues districts in goats, on three different levels:

- Proteomic characterization of different subcutaneous and visceral adipose tissue deposits.
- Comparison between proteomic maps of subcutaneous adipose tissue deposits (sternum and base of tail), visceral adipose tissue deposits (perirenal and omental) and liver.
- Comparison between the omentum proteomic maps of animals fed with different diets (control, fish oil enriched diet, stearic acid enriched diet).

Materials and methods

Samples were obtained from twelve healthy kids whose mothers were fed with either of three different diets (control, A and B) enriched with different fatty acids (Stearic Acid=diet A, Fish Oil=diet B). Five different adipose tissue samples were taken from each animal, equally distributed between controls, diet A and diet B, and immediately snap frozen in liquid nitrogen and stored at -80 °C. Subcutaneous fat was taken from sternum and base of tail depots; visceral fat was taken from perirenal and omental depots. Samples were taken also from the liver in order to use them as reference samples.

First a descriptive analysis was accomplished, using a shotgun approach on the four control animals. Proteins were obtained after precipitation with six volumes of ice-cold acetone, using the same protocol for liver and fat tissues. After tryptic digestion, peptides were fractionated using a strong cation exchange liquid chromatography and the resultant fractions were merged and further fractionated with a reverse phase liquid chromatography and analysed with a tandem mass spectrometry system (ESI-qTOF) (Danielsen *et al.*, 2011).

The four biological replicates from each of the five tissues were searched separately with ProteinPilot 1.0 software; using a confidence level of 95% for protein identifications.

To compare the proteome profiles across the tissue types, we performed a hierarchical clustering analysis using the Ward's linkage method, an agglomerative approach based on the 'error sum of squares' when merging pairs of clusters.

Data handling and analysis were performed using the statistical software package R and for the functional annotation of the identified proteins we used the Blast2GO tool (http://blast2go.org) (Conesa *et al.*, 2005).

In addition a quantitative comparison of the omentum tissues from twelve animals was made, using the iTRAQ labelling technology. All 3 treatment groups were compared (controls, diet A, diet B), Pools of the four control samples were created to be used as reference between different runs, which allowed comparing directly all relative proteome ratios across all twelve animals, divided in the 3 treatment groups (controls, diet A, diet B). 12 samples and 4 pooled control samples were individually labelled with iTRAQ tags and combined in 4 iTRAQ multiplexed experiments prior to fractionation by nano-LC and analyses by tandem mass spectrometry. Data handling and analyses were performed as described above. The statistical analysis is still in progress.

Results

The high-fat content of the adipose tissues did not interfere with the protein extraction protocol and with the MS/MS analysis. The descriptive study allowed the identification of 400-500 proteins in each of the four adipose tissues, with a percentage of spectral coverage between 57 and 63 and 600-700 proteins were identified with spectral coverage between 65 and 68, in liver samples.

In total 1351 proteins were identified across the 5 tissues using a 95% confidence limit threshold. After disregarding the identification that were only observed once, (the so called 'one-hit-wonders') the protein list was reduced to 919 proteins. Of these 402 are exclusively expressed by the adipose tissue and 158 by the liver. The hierarchical clustering analysis showed a clear difference in protein expression patterns, allowing subcutaneous and visceral adipose tissues to be distinguished by their individual proteome expression patterns, and likewise, adipose tissues and liver tissues were clearly distinguishable by their proteome patterns.

Regarding the comparative-quantitative analysis, the software ProteinPilot 1.0 led to the identification of 700-800 proteins in each run, with a spectra coverage between 55.6 and 61.4.

In total, after removing the 'one-hit-wonder' identification, 634 were identified with 95% of confidence. The statistical analysis is still in progress.

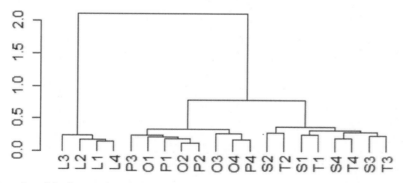

Figure 1. Results of the hierarchical clustering analysis. The distance measure between proteome profiles was calculated as $1-\rho_{ij}$, where ρ_{ij} is the Spearman's rank based correlation between the protein score profiles i and j.

References

Conesa, A., Götz, S., Garcia-Gomez, J.M., Terol, J., Talon M., Robles, M., 2005. Blast2GO: a universal tool for annotation, visualization and analysis in functional genomics research. Bioinformatics 21, 3674-3676.

Danielsen, M., Pedersen, L.J., Bendixen, E., 2011. An in vivo characterization of colostrum protein uptake in porcine gut during early lactation. Journal of Proteomics 74(1), 101-9.

Kershaw, E.E. and Flier, J.S., 2004. Adipose tissue as an endocrine organ. The Journal of Clinical Endoc rinology and Metabolism 89(6), 2548-56.

Lemor, A., Mielenz, M., Altmann, M., von Borell, E., Sauerwein, H., 2010. mRNA abundance of adiponectin and its receptors, leptin and visfatin and of G-protein coupled receptor 41 in five different fat depots from sheep. Journal of Animal Physiology and Animal Nutrition 94(5), e96-101.

Rajesh, R.V., Heo, G.N., Park, M.R., Nam, J.S., Kim, N.K., Yoon, D., Kim, T.H., Lee, H.J., 2010. Proteomic analysis of bovine omental, subcutaneous and intramuscular preadipocytes during in vitro adipogenic differentiation. Comparative Biochemistry and Physiology. Part D, Genomic & Proteomics 5(3), 234-44.

Farm animal proteomics in toxicological studies: useful tool or waste of time? what to learn from rodent studies

T. Serchi[1], I. Miller[2], E. Rijntjes[3], J. Renaut[1], A.J. Murk[3], N.P. Evans[4], E. Ropstad[5], L. Hoffmann[1] and A.C. Gutleb[1]
[1]Centre de Recherche Public – Gabriel Lippmann, Belvaux, Luxembourg; serchi@lippmann.lu
[2]University of Veterinary Medicine, Vienna, Austria
[3]Wageningen University, Wageningen, the Netherlands
[4]University of Glasgow, United Kingdom
[5]Norwegian School of Veterinary Sciences, Oslo, Norway

In recent years sheep (*Ovies aries*) and goats (*Capra aegagrus hircus*) have been increasingly used as model animals in toxicology and a plethora of effects due to exposure with polyhalogenated aromatic hydrocarbons such as polychlorinated biphenyls (PCBs) has been described in these species (1,7-10,12,13,16,17). The effects reported range from the molecular, to the organ and to the whole organism level including reproduction (12,13), behavior (8), stress adaptation (16,17), and bone tissue (7). PCBs have recently even been tested *in vitro* on their potential to induce changes in the protein pattern in exposed cells and a range of significant changes within the proteome of exposed cells has been described (11). To the best of our knowledge no toxicoproteomic studies have been performed on sheep or goat, despite the proven usefulness of the toxicoproteomic approach for toxicological research in other animals (2-5,14).

The experiments and analysis of results for proteomic experiments on the tissues (plasma, liver, hypothalamus, pituitaries, etc.) from experimentally exposed sheep (7-9,16,17) are currently ongoing. We aim, therefore, to show that proteomics can be a valuable tool for toxicological research in a variety of species including farm animals illustrating its potential with available information from a relevant study with rats from a similar experimental paradigm. In so doing we aim to show that proteomic studies have a valuable place in toxicological studies, even those using farm animals, in order to better understand mechanism of toxicity and to make better use of highly valuable tissue material generated from expensive long-term animal experiments.

Brominated flame-retardants and related halogenated aromatic hydrocarbons such as PCBs have been shown to interfere with several hormone systems. Due to the structural resemblance of these compounds to thyroid hormones, in many studies, scientists have focused on thyroid hormone specific endpoints (6). Such hypothesis-driven research is more difficult with compounds with molecular structures that do not resemble known endogenous compounds, such as the brominated flame retardant hexabromo-cyclododecane (HBCD) (Figure 1). Previously performed *in vivo* research with rats has indicated complex effects of HBCD that cannot be explained based on known mechanisms of action (15).

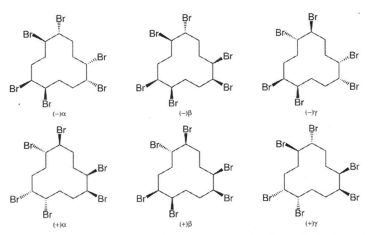

Figure 1. Structures of the main isomers that are present in the technical product of HBCD (http://www.mfe.govt.nz/publications/hazardous/investigation-of-brominated-flame-retardants/html/page11.html).

Therefore the present project followed a proteomic approach to study changes in non-targeted global protein distribution following HBCD exposure. Differentially regulated proteins were studied in liver samples from HBCD exposed rats, and a selection of proteins identified in order to put the results in a toxicological context. The exposed rats were either euthyroid or hypothyroid, and female or male.

Livers of treated rats were collected, ground in liquid nitrogen and proteins extracted in the presence of protease inhibitors. Samples were separated by 2D-DiGE and data evaluated by a multivariate analysis (PCA, hierarchical clustering and multivariate ANOVA). Clear differences were visible between euthyroid and hypothyroid rats (Figure 2), allowing us to select more than 200 differentially expressed spots. The biggest differences were found in relation to gender, highlighting the importance of the choice of gender of experimental animals for toxicological studies.

The complexity of the interactions between cells, genes and the environment, i.e. the interaction of the living matter with the environment, can only be captured by -omics techniques and systems biology that will help to fully understand the complex pattern of induced effects. Proteomics applying either gel-based or non-gel based approaches followed by mass spectrometry including pathway analyses, thanks to its unique holistic *a posteriori approach,* has the potential to elucidate the complex pattern of effects of persistent pollutants in animals used for toxicological studies. As the sample amount necessary for proteomic studies is minute, proteomic analysis does not interfere with the measurement of other parameters. The additional information obtained from already exposed animals clearly contributes to the 3R`s principles (replace, refine, reduce) and our knowledge base and thus proteomics is a highly worthwhile addition to toxicological studies.

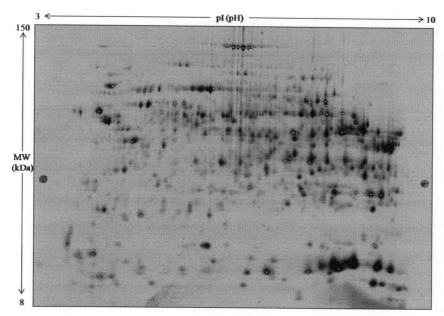

3 ← pI (pH) → 10

150

MW
(kDa)

8

Figure 2. A representative gel of rat liver showing the differentially expressed proteins between euthyroid and hypothyroid rats. Samples were separated by 2D-DiGE on a 24 cm strip (3-10 pH) in first dimension and on a 12.5% polyacrylamide gel in second dimension.

References

1. Berg *et al.*, 2010. Chemosphere, 80, 1144-1150
2. Dorts *et al.*, 2011. Aquat. Toxicol. 103, 1-8
3. Gabrielsen *et al.*, 2011. Aquat. Toxicol. 105, 482-491
4. Gillardin *et al.*, 2009. Mol. Cell Prot. 8, 596-611
5. Gündel *et al.*, 2012. Ecotoxicol. 76, 11-22
6. Gutleb *et al.*, 2010a. Environ. Sci. Tech. 44, 3149-3154
7. Gutleb *et al.*, 2010b. Toxicol. Lett. 192, 126-132
8. Gutleb *et al.*, 2011. Ecotox. Environ. Saf. 41, 1971-2002
9. Kraugerud *et al.*, in press Environ. Toxicol.
10. Krogenaes *et al.*, 2008. Theriogenol. 70, 15-26
11. Lassere *et al.*, 2009. J. Prot. Res. 8, 5485-5496
12. Lyche *et al.*, 2004. Reprod. Toxicol. 19, 87-95
13. Oskam *et al.*, 2005. Reprod. 130, 731-742
14. Silvestre *et al.*, 2010. Sci. Total Environ. 408, 3176-3188
15. van der Ven *et al.*, 2006. Toxicol. Sci. 94, 281-292
16. Zimmer *et al.*, 2008, J. Toxicol. Environ. Health A, 72, 164-172
17. Zimmer *et al.*, in press Environ. Toxicol.

The role of SPEF2 in male fertility

Anu Sironen[1], Jeanette Hansen[2], Bo Thomsen[2], Magnus Andersson[3], Johanna Vilkki[1], Jorma Toppari[4,5] and Noora Kotaja[4]

[1]MTT Agrifood Research Finland, Biotechnology and Food Research, Animal Genomics, FIN-31600 Jokioinen, Finland; anu.sironen@mtt.fi

[2]University of Aarhus, Faculty of Agricultural Sciences, PO Box 50, 8830 Tjele, Denmark

[3]University of Helsinki, Department of Clinical Veterinary Sciences, Helsinki, Finland

[4]University of Turku, Department of Physiology, Finland

[5]Pediatrics, Kiinamyllynkatu 10, 20520 Turku, Finland

The immotile short tail sperm (ISTS) defect is an autosomal recessive disease within the Finnish Yorkshire pig population. The defect is expressed in males as a shorter sperm tail length and immotile spermatozoa. Histological examination of spermatozoa from ISTS affected boars indicates that the axonemal complex and accessory structures of the sperm tail are severely compromised. Cilia in respiratory specimens from ISTS boars are physiologically normal and no adverse effects on reproductive performance of female relatives have been observed suggesting that other ciliated cell types are not influenced. A genetic tool for marker- and gene-assisted selection was needed in order to decrease the severity and economic impact of this defect. Identification and elucidation of the function of a gene crucial for male fertility provides valuable information of events leading to normal sperm formation in all mammals.

In the initial genome wide screen the disease locus was mapped on porcine chromosome 16 within a 3 cM region and a two-marker-haplotype was developed for marker-assisted selection within analyzed families. The disease associated region was further fine-mapped in order to develop a 100% specific test for carrier detection. Sequence analysis of the *SPEF2 (KPL2)* gene revealed the presence of an inserted Line-1 retrotransposon within an intron. The insertion affects splicing of the SPEF2 transcript via skipping of the upstream exon or by causing the inclusion of an intronic sequence, as well as part of the insertion in the transcript. Both changes alter the reading frame leading to premature termination of translation (Sironen *et al.* 2006, 2007). Since 2006 gene assisted selection for ISTS based on this insertion sequence has been made available to pig breeders in Finland.

We have localized the SPEF2 protein during the spermatogenesis in the pig and mouse testis. Yeast two hybrid analysis and co-IP experiments identified a possible interaction partner IFT20. SPEF2 and IFT20 are both localized in the Golgi complex and manchette of elongating spermatids. Furthermore, the presence of SPEF2 in the mature sperm tail midpiece also suggests a structural function for SPEF2 in the sperm tail (Sironen *et al.* 2010).

Our results show that the SPEF2 is important for correct sperm flagella development. Disruption of this process is responsible for the ISTS defect and male infertility in Finnish Yorkshire boars. Expression and interaction studies of the SPEF2 protein allowed the function of SPEF2 to be elucidated. Due to the highly conserved nature of spermatogenesis these results provide novel insights into sperm tail development and male infertility disorders in all mammalian species.

References

Sironen A, Hansen J, Thomsen B, Andersson M, Vilkki J, Toppari J, Kotaja N. Expression of SPEF2 During Mouse Spermatogenesis and Identification of IFT20 as an Interacting Protein. Biol Reprod. 2010 Mar;82(3):580-90.

Sironen A., Thomsen B., Andersson M., Ahola V., Vilkki J. An intronic insertion in *KPL2* results in aberrant splicing and causes the immotile short tail sperm defect in the pig. Proc Natl Acad Sci U S A. 2006 Mar 28;103(13):5006-11.

Sironen A., Vilkki J., Bendixen C., Thomsen B. Infertile Finnish Yorkshire boars carry a full-length LINE-1 retrotransposon within the KPL2 gene. Mol Genet Genomics 2007 Oct;278(4):385-91.

Proteomic analysis of cryoconserved bull sperm to enhance ERCR classification scores of fertility

Alessio Soggiu[1], Cristian Piras[2], Luigi Bonizzi[1], Alessandro Gaviraghi[8], Andrea Galli[8], Hany Ahmed[1], Paolo Sacchetta[3,4,5], Andrea Urbani[6,7] and Paola Roncada[8]

[1]DIPAV, Facoltà di Medicina Veterinaria, Università degli Studi di Milano, Italy; alessio.soggiu@unimi.it

[2]Dipartimento di Scienze Zootecniche, Università degli studi di Sassari, Italy

[3]Centro Studi sull'Invecchiamento (Ce.S.I.), Fondazione 'G. d'Annunzio', Chieti, Italy

[4]Dipartimento di Scienze Biomediche, Università 'G. d'Annunzio' di Chieti e Pescara, Italy

[5]Laboratorio di Enzimologia, Facoltà di Farmacia, Università 'G. d'Annunzio' di Chieti e Pescara, Italy

[6]Dipartimento di Medicina Interna, Università Tor Vergata, Roma, Italy

[7]Fondazione S.Lucia-IRCSS, Rome, Italy

[8]Istituto Sperimentale Italiano L. Spallanzani, Milano, Italy

Introduction

Breeding of dairy cattle for high production and the reproductive management of herd is the biggest problem and it accounts for a large part on costs of production. A negative association has been observed between the level of livestock production and fertility. This is linked both to genetic factors (inbreeding and high production) and physiological factors (metabolic by high production)[1]. A lot of resources have been used for enhancement of cattle fertility but few studies and interventions are reported to control and to enhance the effect on the bull reproductive efficiency. As the patterns of selection and reproductive management of dairy cattle is based on the use of artificial insemination (AI) it is easy to understand the importance of assessing the level of fertility of bull breeder. One method of evaluating relative sire fertility currently used is the estimated relative conception rate (ERCR). ERCR is the difference in conception rate (nonreturn rate at 56 day) of a sire compared with other AI sires used in the same herd[2]. In this work the nonreturn rate was estimated at 56 d for first insemination of lactating cows (www.anafi.it). At present, validation of genomic markers that are able to predict with high confidence high or low fertility of a given sire it is very difficult using population estimates of sire fertility. The reason is because these methods do not measure the bull 'true fertility'[3]. To unravel the biological display of the bull genome, proteomics, that focus at the protein level could lead to the development of novel biomarkers that may allow for detection of bull fertility levels[4,5]. The aim of this study is to evaluate, through the differential proteome analysis, changes in protein expression profiles of spermatozoa from bulls with high fertility (high ERCR score) and low fertility (low ERCR score) in order to identify possible protein markers to be used as indices of fertility.

Methods

Four classes of ERCR score were selected (from very low to very high fertility) for proteomic analysis. Sperm proteins were separated by 2-DE and digitized maps from each class subjected to image analysis with Progenesis SameSpot software. Differentially expressed spots ($P<0.05$) were excised, digested and tryptic peptides analyzed by MALDI-TOF/TOF mass spectrometry.

Results and discussion

Image analysis highlighted three significantly up and down regulated proteins in ERCR groups (Figure 1).

Alpha-enolase was found to be strongly up-regulated in very high fertility (ERCR++) group. In human reproduction, elevated dimeric form of α-enolase (ENO-αα) characterizes abnormal immature spermatozoa, and elevated levels of ENO-S isoform (an isoform sperm-specific[6]) characterizes normally developed spermatozoa[7]. At present there are no data about elevated expression of α-enolase in bull sperm but only in fluid derived from cauda epididymal of mature Hollstein bull in association with a high fertility profile[8,9]. Other two proteins, isocitrate dehydrogenase subunit alpha (IDH-α) and triosephosphate isomerase (TPI) showed highest expression in ERCR-/- group which is associated with a very low score of fertility. Triosephosphate isomerase (TPI), an important glycolytic enzyme, is underexpressed in ERCR++ group respect to the others groups ($P=0.048$). In literature are not present data about

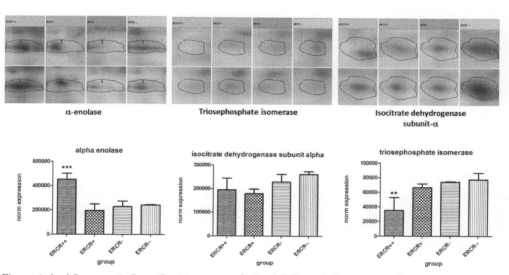

Figure 1. (up) Progenesis SameSpot image analysis of differentially expressed bull sperm proteins, (bottom) normalized expression levels of differentially expressed proteins in the four classes analyzed. **=P<0.01; ***=P<0.001.

this protein in bull sperm but similar profiles of expression are found in human sperm from asthenozoospermic (low motility sperm) patients[10]. In this work there is an overexpression of IDH-α in sperm samples with low ERCR score. The explanation of this phenomenon can be found either in a possible modulation of hypoxia-inducible factor-1 in sperm cell due to several type of metabolic problems[11] or an increased necessity of NADPH in response to an increased oxidative stress. The latter possibility is largely supported by literature data. In effect defective human spermatozoa show intense redox activity and oxidative stress has been associated with impaired sperm motility[12] and also sperm-oocyte fusion is inhibited by oxidative stress[13].

Conclusion

In conclusion, the present study provides the first evidence for protein variations linked at the ERCR values in the bull sperm proteome and demonstrates that 2-D gel electrophoresis coupled to mass spectrometry and bioinformatics is useful for the identification of biomarkers for evaluation the level of fertility. The present data have indicated several possible candidate protein biomarkers for high and low ERCR. Further investigations will be necessary to evaluate possible use of these markers in fast screening of bull semen (by flow cytometry), and to clarify the causes of bull infertility.

Acknowledgements

Work supported by PRO.ZOO Project. ISILS.

References

1. Bach, A., Valls, N., Solans, A. & Torrent, T. Associations between nondietary factors and dairy herd performance. *J Dairy Sci* 91, 3259-67 (2008).
2. Clay, J.S. & McDaniel, B.T. Computing mating bull fertility from DHI nonreturn data. *J Dairy Sci* 84, 1238-45 (2001).
3. Amann, R.P. & Dejarnette, J.M. Impact of genomic selection of AI dairy sires on their likely utilization and methods to estimate fertility: a paradigm shift. *Theriogenology* (2011).
4. Tomar, A.K. *et al*. Differential proteomics of sperm: insights, challenges and future prospects. *Biomark Med* 4, 905-10 (2010).
5. Gaviraghi, A. *et al*. Proteomics to investigate fertility in bulls. *Vet Res Commun* 34 Suppl 1, S33-6 (2010).
6. Edwards, Y.H. & Grootegoed, J.A. A sperm-specific enolase. *J Reprod Fertil* 68, 305-10 (1983).
7. Martinez-Heredia, J., de Mateo, S., Vidal-Taboada, J.M., Ballesca, J.L. & Oliva, R. Identification of proteomic differences in asthenozoospermic sperm samples. *Hum Reprod* 23, 783-91 (2008).
8. Moura, A.A., Souza, C.E., Stanley, B.A., Chapman, D.A. & Killian, G.J. Proteomics of cauda epididymal fluid from mature Holstein bulls. *J Proteomics* 73, 2006-20 (2010).
9. Moura, A.A., Chapman, D.A., Koc, H. & Killian, G.J. Proteins of the cauda epididymal fluid associated with fertility of mature dairy bulls. *J Androl* 27, 534-41 (2006).

10. Zhao, C. *et al*. Identification of several proteins involved in regulation of sperm motility by proteomic analysis. *Fertil Steril* 87, 436-8 (2007).

11. Paul, C., Teng, S. & Saunders, P.T. A single, mild, transient scrotal heat stress causes hypoxia and oxidative stress in mouse testes, which induces germ cell death. *Biol Reprod* 80, 913-9 (2009).

12. Aitken, R.J. & Baker, M.A. Oxidative stress and male reproductive biology. *Reprod Fertil Dev* 16, 581-8 (2004).

13. Baker, M.A. & Aitken, R.J. The importance of redox regulated pathways in sperm cell biology. *Mol Cell Endocrinol* 216, 47-54 (2004).

Proteomics biomarkers for pork quality

Marinus F.W. te Pas[1], Mariusz Pierzchala[2], Els Keuning[1], Ron E. Crump[1], Sjef J.A. Boeren[3], Maria Hortos[4], Marina Gisbert[4], Alejandro Diestre[5] and Leo Kruijt[1]
[1]*Animal Breeding and Genomics Centre, Wageningen UR Livestock Research, the Netherlands; marinus.tepas@wur.nl*
[2] *Institute of Genetics and Animal Breeding, Polish Academy of Animal Sciences, Jastrzębiec, Poland*
[3]*Laboratory of Biochemistry, Wageningen University, Wageningen, the Netherlands*
[4] *IRTA, Monells, Spain*
[5] *PIC, Spain.*

Abstract

Proteome expression profiles can be used as biomarkers to predict traits with low heritability and large environmental influences. Porcine meat quality typically is such a trait. Proteome profiles of 150 meat samples were determined and associated with measured technical meat quality traits. Biomarkers tests were developed and their predictive potential determined. Differentially expressed proteins related to variation in drip loss and ultimate pH were determined and the underlying biological mechanism evaluated with bioinformatics. The biological mechanism also explains the connections between these two meat quality traits.

Introduction

The proteome of tissues is an important predictor of the phenotype of traits, especially if the heritability of the trait is low. Genetic markers are poor predictors of such traits due to the large influence of the environment on the trait. Biomarkers measuring the expression of the genome at the proteome level determine the genetic capacity and the interaction with the environment. Thus, proteomic biomarkers can be used to predict such traits. Meat quality is typically such a trait with heritability as low as 10-15%. Meat quality is a complex trait and composed of technical properties and consumer eating perception. Meat quality is influenced by feeding, housing, management, transportation, etc. of the animals.

The objectives of this study were (1) to identify associations between the proteome profiles of 150 pork samples and technical meat quality traits measured in these samples; (2) to identify proteins and underlying biological mechanisms explaining variation in meat quality traits; and (3) to show the potential of proteomic biomarkers to predict meat quality traits.

Materials and methods

Longissimus muscle samples of 150 Large White x Duroc sows and castrates were used to isolate the sarcomere proteome. Proteome profiles were determined using SELDI-TOF equipment and the CM10, Q10, and IMAC30 arrays. Associations between proteome peaks and meat quality traits were determined using SAS and Partial least squares regression (PLSR) of adjusted phenotypes was used to investigate the predictive ability of protein biomarkers and combinations of protein biomarkers for meat quality traits. The predictive capacity of biomarkers were evaluated by separating the samples into six groups. Five were used to determine the association again and these associations were used to predict the meat quality of the samples in group six based upon the expression of the proteins in these samples. Fourier transform mass spectrometry (FTMS) analyses were performed to identify the differently expressed proteins underlying the variation of the meat quality traits for drip loss and ultimate pH. High and low trait values were compared with average trait values to determine differentially expressed proteins. Bioinformatics was used to study the underlying biological mechanisms.

Results and discussion

Biomarker test development

Table 1 shows the association between meat quality traits and proteomic peaks as determined by SELDI-TOF using a linear model in the PLSR. Several traits showed multiple peaks with

Table 1. Results of linear protein effect selection based on variable importance in projection in PLSR. Cross-validated number of latent components (Nc) from the model which resulted in the lowest PRESS statistic and the proteins included in this model. Peaks are indicated by their m/z value and array type(I = IMAC30; C = CM10; Q = Q10).

Trait	Nc	PRESS	Selected proteins	
			N	Proteins
Drip loss	2	16.06	5	C06624_7, C08453_2, I05698_9, Q01350_8, Q02622_1
Fat34FOM	8	1,199.32	5	C03140_9, C06624_7, I03904_9, I04417_7, I06617_2
Loin34FOM	2	3,005.84	3	C03140_9, C08453_2, I06617_2
IMF	4	136.94	1	I04417_7
Fatthickham	6	2,402.63	3	C08453_2, I04417_7, I06617_2
NPPCmarbling	7	87.32	7	C03140_9, C03905_2, C06110_0, C06624_7, C08453_2, I03904_9, I04417_7
NPPCcolor	2	63.07	5	C04419_7, C05001_4, C06624_7, C08453_2, C010260_
Ultimate pH LD	9	1.90	9	C04419_7, C05702_9, C06624_7, C08453_2, I03904_9, I04417_7, I06617_2, Q01350_8, Q02622_1
Ultimate pH SM	4	2.20	20	C03140_9, C03612_5, C03905_2, C03975_1, C04419_7, C04988_6, C05620_4, C05702_9, C06110_0, C06624_7, C08453_2, C08478_3, C010260_, I03904_9, I04417_7, I06617_2, I08447_3, Q01288_2, Q01350_8, Q02622_1

association. Using a quadratic model more peaks were identified (data not shown). It should be noticed that peaks with similar m/z values on different array types usually are different proteins.

The accuracy of the prediction of combinations of proteins is shown in Table 2. The results show that proteomic biomarkers can predict meat quality traits with up to 80% accuracy.

The number of peaks necessary for this accuracy varies from three up to more than ten. Thus, such biomarker test may require well-equipped laboratories to measure the biomarkers. Alternatively novel methods such as multiple dipstick methods need to be developed for practical use of the biomarkers on-site and in-line.

Table 2. Accuracy of prediction.

Trait	Mean	s.d.	Min	Max
Drip loss	0.481	0.148	0.181	0.800
Fat34FOM	0.466	0.161	0.140	0.720
Loin34FOM	0.202	0.232	-0.276	0.678
IMF	0.285	0.176	-0.088	0.593
Fatthickham	0.374	0.168	0.064	0.609
NPPCmarbling	0.292	0.194	-0.181	0.728
NPPCcolor	0.164	0.148	-0.117	0.511
Ultimate pH LD	0.382	0.221	-0.152	0.695
Ultimate pH SM	0.515	0.160	0.069	0.841

Biological mechanisms determining drip loss and ultimate pH

Differentially expressed proteins in high and low drip loss and ultimate pH as compared with medium trait values were identified. In total 26 proteins were differently expressed in drip loss, and 22 for ultimate pH. Table 3 shows the results of the bioinformatics analysis focusing on biological activities of the proteins. It should be noted that several of the proteins showed more than one biological activity, and eight proteins were found in more than one comparison. Cross effects between drip loss and ultimate pH were mainly found among the energy metabolism, protein degradation and calcium storage mechanisms, suggesting these mechanisms as central molecular biological mechanisms connecting these traits.

Summarizing, we developed a predictive set of proteomics biomarkers for meat quality traits with low heritability, and detected possible biological mechanisms underlying these traits.

Table 3. Number of regulated proteins per biological activity for each trait value.

Biological activities	Drip loss				Ultimate pH			
	High		Low		High		Low	
	Up	Down	Up	Down	Up	Down	Up	Down
Energy metabolism	4	2	3	3	2	1	3	1
Protein degradation	1	3		1	1	1	1	
ECM	1			1	1			1
Signal transduction	1	2		4	2			1
Chaperonin (structural)	1							
Muscle structural protein		3		1	2	1		
Calcium metabolism		2		3	2			1
Apoptosis		1		1		1		1
Nucleotide metabolism		1		1	2			
Muscle mass determination				1				1
Anti-oxidant				2		1		
HSP	1	2		1				

Acknowledgements

The authors gratefully acknowledge financial participation from the European Community under the Sixth Framework Programme for Research, Technological Development and Demonstration Activities, for the Integrated Project Q-PORKCHAINS FOOD-CT-2007-036245, and from the 'Kennisbasis' (Knowledge Base) grant no KB05-003-02 of the Dutch Ministry of Agriculture, Nature and Food security.

Proteomics as a tool to understand fish stress in aquaculture

Tomé S. Silva[1,2], Pedro M. Rodrigues[1], Elisabete Matos[1], Tune Wulff[2], Odete D. Cordeiro[1], Ricardo N. Alves[1], Nadège Richard[1], Mahaut de Vareilles[1], Flemming Jessen[2], Jorge P. Dias[1] and Luís E.C. Conceição[1]
[1]CCMAR – Centro de Ciências do Mar do Algarve, Faro, Portugal; tome@tomesilva.com
[2]DTU Food – Fødevareinstituttet, Kgs. Lyngby, Denmark

Physiological and psychological stress are important concerns in economic sectors that rely on animal husbandry, particularly in finfish aquaculture, as impaired fish welfare often implies worse fish health, productivity and final product quality traits. Moreover, this also brings obvious ethical implications.

Stress itself is a very complex and difficult to define concept from a quantitative point of view, as there is no universally-accepted objective definition of *stress* or *animal welfare*. Ashley defines animal welfare as the freedom from hunger and thirst, discomfort, pain, injury, disease, fear and distress, and the freedom to express normal behavior [1], which is a fairly encompassing view that underlines the multifactorial nature of welfare. In the context of aquaculture, fish can be seen as being often exposed to several distinct sources of physiological and psychological stress (Figure 1), more or less avoidable, and that can vary in terms of type, intensity and duration.

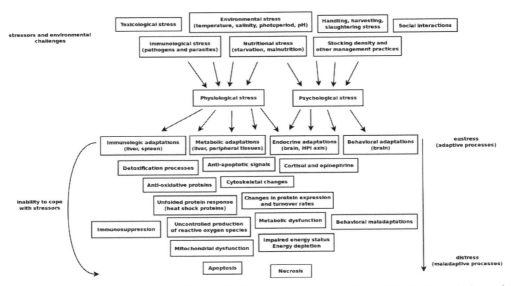

Figure 1. Global overview of the relationship between stressors and physiological/psychological/cellular stress.

Currently, the most accepted view on physiological and psychological stress has shifted from the classic 'homeostatic' model, where organisms are seen as continuously striving towards a static (conceptually optimal) state, to an 'allostatic' model, that more accurately accounts for the adaptive and dynamic nature of biological systems [2]. This model assumes that understimulation (due to lack of signals and challenges from the environment) can be as detrimental to fish welfare and development as overstimulation. Also, it distinguishes adaptive changes ('eustress') from maladaptive changes ('distress'), where the latter usually occur when the intensity or duration of the challenges overcome the organism's coping capacities.

Although fish welfare studies have usually relied on classical physiological measurements of stress (e.g. plasma cortisol, glucose and lactate), it is increasingly clear that targeting signals of cellular stress, particularly using holistic approaches like proteomics, provides complementary information on biological state of fish that enables a deeper understanding of the relation between fish welfare, physiological/psychological stress and cellular stress. Even so, obtaining universally valid quantitative estimators of *stress* or *welfare* status and clearly distinguishing between adaptive and maladaptive processes are far from trivial tasks due to contextual and inter-individual differences in the way fish react to stress. This underlines the need to integrate these different complementary signals of physiological, psychological and cellular stress to have a more clear insight on stress response processes.

As the biological response to physiological and psychological stress affects several organs and tissues in fish, there is a wide range of tissues that are usually targeted by stress studies. Our research group mainly focuses on liver (due to its central role in metabolic control and detoxification) and skeletal muscle (since it is the most important part of the fish, from an aquaculture production and fish quality point of view), and it is interesting to note the overlaps and differences when comparing the effect of physiologic stress on the two tissues, using the same biological model (e.g. gilthead seabream): while physiological stress induces in both tissues proteomic changes at the level of cytoskeletal and redox homeostasis systems, hepatic changes usually involve deep changes in carbohydrate, lipid and amino acid metabolism [3], while skeletal muscle is characterized more by proteomic changes involving energy storage, mobilization and transfer processes (e.g. glycogen phosphorylase and proteins of the phosphotransfer systems) and proteins related to the ubiquitin/proteasome pathways.

Comparing the effect of similar stressors (e.g. handling stress) on the hepatic proteome of different fish species (e.g. gilthead seabream [3] *vs.* Senegalese sole [4]), we observed a significant overlap in terms of affected pathways (mostly cytoskeleton, primary metabolism, redox homeostasis and detoxification processes), but the overlap in terms of specifically affected proteins is quite weak. Also, it is relevant to note the proteomic similarities even between very distinct *in vivo* and *in vitro* models of stress: results from subjecting rainbow trout fibroblasts to prolonged anoxia [5] show similar cellular responses to stress to the ones observed *in vivo* (regulation of proteins involved in cytoskeleton, energy metabolism and redox proteins), reinforcing the notion of a universal basis in the mechanisms of stress response.

Additional studies with these in vitro models shows, nonetheless, that even applying two very similar stressors to a population of very identical experimental specimens, in a very controlled environmental context, still leads to slightly different cellular responses to stress, in terms of specifically affected proteins [6].

Given this, it is again wise to remember the importance of biological context and complementary information to obtain a clear picture of the physiological state of fish. Recently, during an experiment meant to test the effect of a dietary treatment on gilthead seabream skeletal muscle, we obtained proteomic results that could be interpreted as a sign of mild nutritionally-induced stress (some increased heat shock protein and aldehyde dehydrogenase isoforms). On the other hand, all other measured parameters implied improved physiological state in fish subjected to the dietary treatment (higher fillet yield, improved muscle energy status, reduced signs of oxidative stress and mostly unchanged signs of proteolytic potential), pointing away from a distress state. As such, we underline the importance of taking a holistic strategy when approaching the topic of fish welfare, integrating classical signals of stress with more untargeted information on cellular processes (from -omics techniques, for instance) to ensure biological signals are interpreted in a truthful and consistent way.

Proteomic techniques are important tools in understanding stress in aquaculture, not only due to their generally untargeted strategies, but also because this subject can only be successfully understood if a multidisciplinary approach is undertaken, taking into account the central role of proteins in metabolism. In this sense, it is also true that correct interpretation of proteomic results in the context of stress requires not only co-measurement context, but also experimental context and the use of experimental designs that take into account the impact of several factors (e.g. developmental and seasonal changes, environmental factors, prior nutritional and physiologic state) on measured stress response signals. It is therefore clear that an integrative systems biology approach is central to adjust aquaculture management practices so that fish welfare and health is ensured, and that proteomics provides key information here, at the level of cellular stress responses.

References

1. Ashley P.J. (2007). Fish welfare: Current issues in aquaculture. Applied Animal Behaviour Science, 104(3-4): 199-235.
2. Martins C.I.M., Galhardo L., Noble C., Damsgård B., Spedicato M.T., Zupa W., Beauchaud M., Kulczykowska E., Massabuau J.C., Carter T., *et al.* (2011). Behavioural indicators of welfare in farmed fish. Fish Physiology and Biochemistry, DOI 10.1007/s10695-011-9518-8.
3. Alves R.N., Cordeiro O.D., Silva T.S., Richard N., de Vareilles M., Marino G., Di Marco P., Rodrigues P.M., Conceição L.E.C. (2010). Metabolic molecular indicators of chronic stress in gilthead seabream (*Sparus aurata*) using comparative proteomics. Aquaculture, 299: 57-66.

4.   Cordeiro O.D., Silva T.S., Alves R.N., Costas B., Wulff T., Richard N., de Vareilles M., Conceição L.E.C., Rodrigues P.M. (2012). Changes in liver proteome expression of Senegalese sole (*Solea senegalensis*) in response to repeated handling stress. Marine Biotechnology, in press.

5.   Wulff T., Jessen F., Roepstorff P., Hoffmann E.K. (2008). Long term anoxia in rainbow trout investigated by 2-DE and MS/MS. Proteomics, 8: 1009-1018.

6.   Wulff T., Hoffmann E.K., Roepstorff P., Jessen F. (2008). Comparison of two anoxia models in rainbow trout cells by a 2-DE and MS/MS-based proteome approach. Proteomics, 8: 2035-2044.

A peptidomic approach to biomarker discovery for bovine mastitis

Rozaihan Mansor[1], William Mullen[2], David C Barrett[1], Andrew Biggs[3] Amaya Albalat[2], Justyna Siwy[4], Harald Mischak [2,4] and P. David Eckersall[1]
[1]*Institute of Infection, Immunity & Inflammation & School of Veterinary Medicine, University of Glasgow, Glasgow, United Kingdom; r.mansor.1@research.gla.ac.uk*
[2]*Biomarkers & Systems Medicine, University of Glasgow, Glasgow, United Kingdom*
[3]*The Vale Veterinary Laboratory, Station Road Tiverton Devon, United Kingdom*
[4]*Mosaique-Diagnostics, Hannover, Germany*

Introduction

Mastitis is an inflammation of mammary gland parenchyma which is characterized by a range of physical and chemical changes of milk and pathological changes in the udder tissues (Radostits *et al.*, 2000). Significant milk changes that can be observed in bovine mastitis are the presence of clots in milk, milk discolouration and high levels of leukocyte numbers in affected milk. Clinical signs in bovine mastitis comprise heat, pain and swelling of the udder. Mastitis is usually caused by bacterial infection with major pathogens such as *Streptococcus agalactiae* and *Staphylococcus aureus* as well as environmental pathogens, including *Streptococcus* species and environmental coliforms (Gram negative bacteria *Escherichia coli*).

Mastitis is one of the most prevalent diseases in dairy cows. It is an endemic disease and one of the most costly diseases affecting dairy herds. It affects the quality of the milk through changes in milk composition and thus, has economic consequence for the entire dairy industry. Discarded milk during treatment as well as a persistent decrease in milk production is the main detrimental effect that contributes to the economic impact of mastitis and can be caused by both clinical and subclinical mastitis. While identification of clinical mastitis can be achieved by identification of the clinical signs described above, subclinical mastitis where these signs are not evident is a more problematic diagnostic task. Currently the industry standard method of quantifying subclinical mastitis is by measurement of somatic cell counts (SCC) in milk which requires submission of a sample to a laboratory. Identification of alternative biomarkers of mastitis which could be adapted to rapid and accurate, on-farm diagnostic systems would be a valuable tool for early detection and treatment of mastitis. This would be particularly useful if the causative organism could be identified allowing better targeted antimicrobial treatment.

The analysis of changes in the peptide content of biological fluid such as serum or urine is increasingly being used as a means to discover biomarkers for the diagnosis and monitoring of disease. There have been a number of studies, especially using urine, that have developed biomarkers models for a number of different diseases. Those investigated so far range from renal disorders (Metzger *et al.*, 2010) to cardiovascular disease (Delles *et al.*, 2010) and diabetes (Maahs *et al.*, 2010).

It would seem likely that changing patterns of the peptides of milk could provide a source of biomarkers for mastitis such that application of advanced peptide biomarker detection and analysis could be a valuable addition to the diagnosis of mastitis.

The aim of this investigation was to undertake a study of the peptides present in milk from healthy dairy cows in comparison to milk from cows with naturally occurring clinical mastitis to determine if peptide biomarkers for the disease could be identified. A second aim was to determine if peptide biomarkers could be identified which could distinguish between the bacteria causing the mastitis by analysis of samples in which either *E.coli* or *S. aureus* had been identified as the pathogenic agent.

Material and methods

Milk samples from clinical mastitic cases were obtained from Vale Veterinary Laboratory, Devon, UK. Bacteriological culture was performed in the Bacteriology Laboratory of Vale Veterinary Laboratory to determine the bacterial cause of bovine mastitis. Milk samples from mastitis caused by either *E. coli* (n=10) or *S. aureus* (n=6) infection were selected for further analysis. Control milk samples (n=10) from healthy cows from the University of Glasgow dairy herd were used in this study as comparison to those clinical mastitic cases. The healthy milk samples were confirmed by somatic cell counts being <100,000 cells/ml.

The samples were filtered through a 20 kD filter prior to being desalted on a PD 10 column. Samples were lyophilized and stored prior to analysis when they were re-suspended in HPLC-grade H_2O shortly before CE-MS analyses using a P/ACE MDQ capillary electrophoresis system (Beckman Coulter, Fullerton, USA) on-line coupled to a micrOTOF MS (Bruker Daltonic, Bremen, Germany).

Result and discussion

Initially in this work we investigated whether differences existed in the milk proteome of healthy cows (n=10) when compared to mastitic cows, regardless of their bacterial cause (n=27). The polypeptide fingerprints of these groups were very different as shown in Figure 1. The next step was to investigate the differences in milk proteome of infected samples caused by two different bacterial pathogens (*E. coli* and *S. aureus*). The polypeptide profiles of these 2 different groups did not differ as markedly as can be seen in Figure 2.

In the first model, we selected discriminatory peptides that were able to differentiate between the control and infected groups with the area under curve (AUC) set at more than 0.99, Benjamini-Hochberg (BH) was set at less than 0.05 and Tmax less than 0.05. Following this

Control

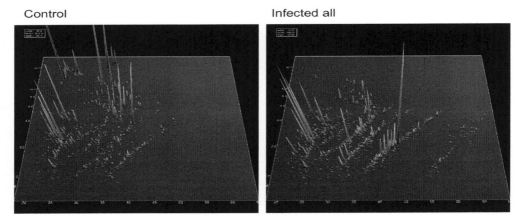

Figure 1. Peptide fingerprints from control and infected milk.

E.Coli   S. Aureus

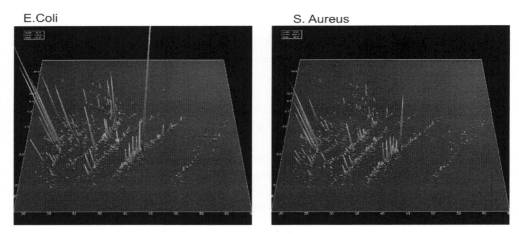

Figure 2. Peptide fingerprints from Escherichia coli *and* Staphylococcus aureus *infected milk.*

very stringent conditions a group of 31 peaks were identified that can be used to discriminate between control and infected groups. From the statistical analysis, this model possessed 100% value for its sensitivity, specificity and accuracy.

Based on the polypeptide fingerprints described above, multiple testing of statistical analysis was done to discriminate peaks between two infected groups caused by *E. coli* and *S. aureus*. The model chosen had an area under curve (AUC) set at more than 0.99, Benjamini-Hochberg (BH) was set at less than 0.05 and Tmax less than 0.05, 6 peaks were identified that can be used to discriminate between *E. coli* and *S. aureus* groups. The sensitivity value for this model was 100% whilst the specificity value was 91.67% and the accuracy value was 96.3%.

Our results indicate that CE-MS proteomics can be used to identify milk from cows with clinical mastitis and to distinguish between milk infected by *E. coli* and *S. aureus* bacteria. Future investigation is required to determine if subclinical mastitis can be similarly identified.

Acknowledgements

Support for RM from the Malaysian Ministry of Higher Education is gratefully acknowledged.

References

Delles, C. *et al*, (2010) Urinary proteomic diagnosis of coronary artery disease: identification and clinical validation in 623 individuals. *J Hypertens* 28, (11), 2316-22.

Maahs, D.M. *et al*, (2010) Urinary collagen fragments are significantly altered in diabetes: a link to pathophysiology. *Plos One* 5, (9).

Metzger, J. *et al*, (2010) Urinary excretion of twenty peptides forms an early and accurate diagnostic pattern of acute kidney injury. *Kidney International* 78, (12), 1252-1262.

Radostits, O.M. *et al*,. (2000). Veterinary Medicine. 9th Edn, ELBS-Balliere Tindal, London pp:563-618.

Acknowledgements

The Organizing Committee of the Vilamoura 2012 - Farm Animal Proteomics meeting acknowledges the support of the entities and individuals that made possible this publication and event.

Major sponsors

ThermoUnicam (www.thermounicam.pt)

VWR (www.vwr.com)

Bio-rad (www.bio-rad.com)

Organizing support

CCMAR – Centre of Marine Sciences (www.ccmar.ualg.pt)

Universidade do Algarve (www.ualg.pt)

COST Action FA1002: Farm Animal Proteomics (www.cost-faproteomics.org/)

COST / European Science Foundation (www.cost.esf.org)

Individual contributions

André Martinho de Almeida (IICT, Lisboa, Portugal)
Andreia Pinto (CCMAR, Universidade do Algarve)
Cristina Inácio (CCMAR, Universidade do Algarve)
David Eckersall (University of Glasgow)
Jillian Bryce (University of Glasgow)
Pedro M. Leal Rodrigues (CCMAR, Universidade do Algarve)

Printed in the United States
by Baker & Taylor Publisher Services